Historical Fighting Fundamentals

German Longsword
by
Kyle Griswold

Copyright 2022, All Rights Reserved

ISBN: 978-1-956090-02-4

EDITOR:
QUINN RICHARDSON

PHOTOGRAPHY:
JON MURRAY

TRANSLATION:
STEPHEN CHENEY

MODELS:
BRITTANY REEVES
KYLE GRISWOLD
SHANE GIBSON

INTERIOR ILLUSTRATIONS:
DYLAN SMITH

COVER ILLUSTRATION:
MARIANA LOPEZ

ADDITIONAL THANKS:
BRITTANY REEVES
SEAN FRANKLIN
RICHARD MARSDEN
JOHN KNOCH
MICHAEL COFFEY

BROKEN BLADE
P U B L I S H I N G

Mordhau Historical Combat

TABLE OF CONTENTS

PREFACE

My goal when writing this book was to provide a basic foundation for understanding the German tradition attributed the fencing master, Johannes Liechtenauer. This art has been reconstructed through the study and translation of historical manuals. Due to the nature of this revived martial system, modern practitioners do not universally agree on its translation or interpretation. I have attempted to present the actions and concepts within this book in a manner which adheres to the various texts from which they originate.

Having owned and operated a successful school for several years, I understand that new Kunst des Fechtens (KdF) students can sometimes be overwhelmed with the material presented within the glosses and uncertain about which manuscripts to focus on. While it is highly encouraged for students to go directly to the source material and attempt to decipher its meaning, it is my hope that this book will be a helpful guide to explain the core concepts within the KdF system. This book also offers mechanical explanations for how certain actions would likely be performed. The reader may or may not agree with some of these interpretations, as some are not expressly defined within the glosses. I believe this advice, at the very least, complies with the glosses and has proven to function.

I attempted to present most of the information within this book in the most objective way possible. Techniques and concepts are explained in a manner which I believe directly reflects the various historical texts where they originate. Information which is more subjective in nature appears within this book in gray sidebars. This supplemental text reflects my own interpretations and beliefs which can not be supported by the glosses in absolute terms.

Readers should challenge the ideas and interpretations within this book. Test them. Rigorously. Do not allow your journey into this art to end with the conclusion of this reading. What is presented here is only a glimpse into Liechtenauer's art. Seek out other material, other instructors, and conflicting interpretations; drill, spar, and compete against others. Perhaps most importantly, go back to the historical texts and test everything you are doing to ensure it complies with spirit of this art.

-Kyle Griswold

Historical Context

Historical European Martial Arts (HEMA) is a modern umbrella term which describes a number of martial arts which were preserved with historical manuscripts. HEMA itself is not a system of fighting, and therefore is not a universal martial art with by which Europeans performed armed or unarmed combat. HEMA sources typically documented dueling and self-defense in a civilian setting.

I have chosen to define the HEMA tradition presented within this book as *Kunst des Fechtens* (KdF), or the Art of Fighting. For the purpose of this book, KdF refers to the specific martial art attributed to the German fencing master, Johannes Liechtenauer. Some may take umbrage with my choice to identify Liechtenauer's art with a term which could define all contemporary Germanic martial arts, though I believe using it in this manner will not confuse the reader.

Liechtenauer's art was preserved with a cryptic poem known as the *Zettel* (Recital). The Zettel is intentionally obscure and impossible to interpret without the glosses (commentaries) written by later fencing masters. Many of these glosses are contained within 15th and 16th century fencing manuals called *fechtbuchs* (fight books).

The glosses are not identical and occasionally conflict with one another. Later fencing masters, such as Joachim Meyer, presented material which at times appears very different from that found in the earlier glosses. Most of the information expressed within this book reflects the writing found in the early KdF glosses. The specific texts which this book focuses on are those of the RDL: a family of glosses attributed to the masters Sigmund Ringeck, Peter von Danzig, and Jud Lew. While early KdF is the primary focus of this book, other masters will be referenced when appropriate.

Joachim Meyer 1570

Hans Talhoffer 1467

Longsword Fundamentals

The *haupstucke* (chief pieces) are the 17 principal techniques and concepts which govern fencing with the longsword. The application and meaning of the haupstucke is found within the various plays contained in the glosses. The haupstucke appear within the Zettel in the following order:

- Zornhau – Wrath Cut
- Krumphau – Crooked Cut
- Zwerchau – Thwart Cut
- Schielhau – Squinting Cut
- Scheitelhau – Parting Cut
- Vier Leger – Four Positions
- Vier Versetzen – Four Displacements
- Nachreissen – Racing-after
- Uberlauffen – Running-over
- Absetzen – Setting-off
- Durwechseln – Changing-through
- Zucken – Twitching/Pulling
- Durchlauffen – Running-through
- Abschneiden – Slicing-off
- Hend Trucken – Hand-pressing
- Hengen – Hanging
- Winden – Winding

Before this book delves into the haupstucke, the reader will be presented with the common teachings found in the glosses. This chapter will also provide the reader with mechanical fundamentals which contemporary readers of the glosses would have been presumably familiar with. Any advice presented here that is not explicitly detailed in the historical texts will be supported with other evidence.

The Longsword

When used with two hands held predominantly upon the grip, swords within the KdF tradition are called longswords. When held with one hand gripping the blade, the same sword would be called a short or half-sword. Though possessing some common properties, longswords vary considerably in length, weight, and appearance.

Longswords possess a hilt which is further divided into the grip, crossguard, and pommel. The weapon has two edges. When held, the edge facing away from the wielder is known as the long or true edge, while the edge facing toward the wielder is called the false or short edge. The length of the blade nearest the hilt is referred to as the strong, and the length of blade nearest the point is the weak.

Federschwerts (feather swords), or feders, are historical training swords. These trainers have a rebated and narrowed blade. The widened section of the blade near the hilt is known as the *schilt* (shield).

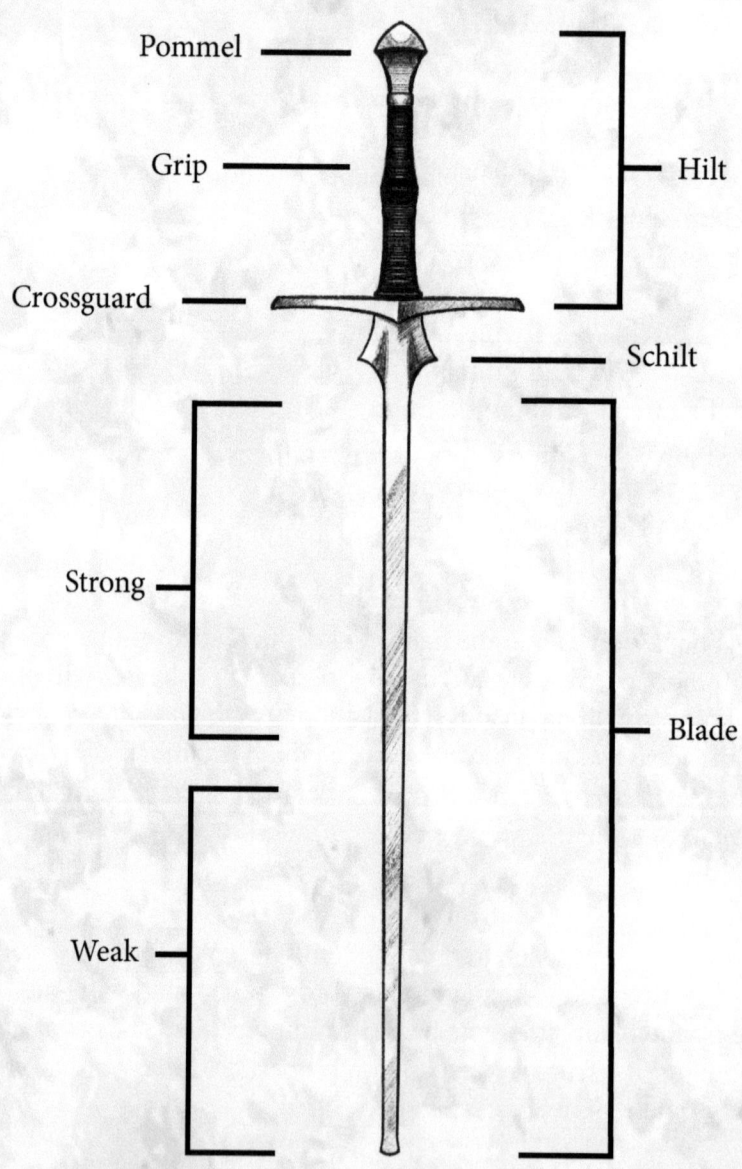

Zufechten (Onset) / Krieg (War)

The Zettel advises that all arts have *measure*, referring to the varying distances in which combatants engage one another. When combatants are maneuvering and preparing to strike one another, they are in *zufechten,* or the onset. When combatants enter the measure necessary to strike their opponent, they are within the *krieg,* or war.

> The above definitions are perhaps overly simplified. Whether the onset is any distance of fighting where fencers cannot reach one another or something more specific is uncertain. When attempting to enter the krieg from the onset, most attacks are performed with extended arms and a step; which is necessary to close the distance.

Vor / Nach / Indes

Fencers should attempt to take the initiative when fighting and look for opportunities to attack their opponent rather than wait and defend. When one fencer strikes before their opponent, they are in the *vor* or before; this is also referred to as seizing the vor. The initiating strike or, *vorschlag* (fore-strike), should threaten the target so they must defend against it or be struck. Whether the vorschlag hits or not, the attacker should seek to remain within the vor, so that they continuing to threaten their target with a series of *nachslags* (after-strikes) once they have entered the krieg. Ideally, the attacker forces their opponent to constantly defend, denying them an opportunity to carry out their own attacks.

When forced to defend against an attack, the fencer is considered to be in the *nach* (after). The danger of remaining in the nach is that the fencer would be thus forced to continue chasing after their opponent's weapon, and never take control of the fight. Therefore, it is important that the defender quickly seeks to turn their defense into an attack. When the defender becomes the attacker, they seize the vor away from their opponent.

Whether working from the vor or nach, a fencer should not rush recklessly from one action to the next. All fencing should be performed in *indes* (in-the-moment), based on the strong and weak of the sword. When blades touch, they bind upon one another, if even for a moment. The strong of the blade can resist pressure in the bind while the weak of the blade yields to pressure. Fencers should rely on pressure to make decisions in-the-moment, rather than flying off of the bind too soon, or remaining in the bind unnecessarily.

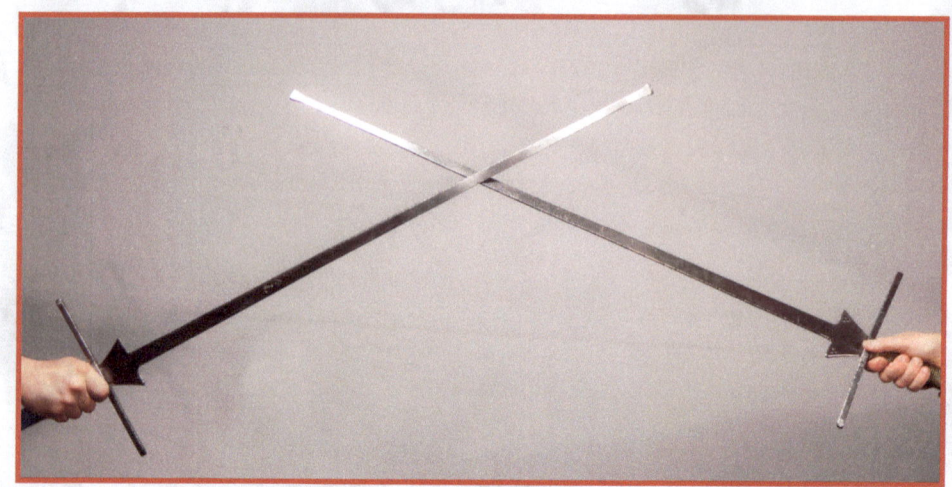

Weak on weak.

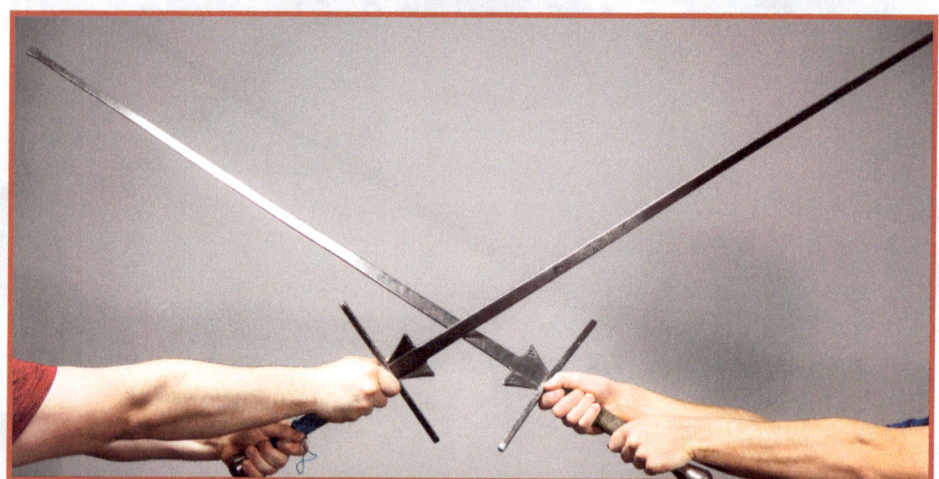

Strong on strong.

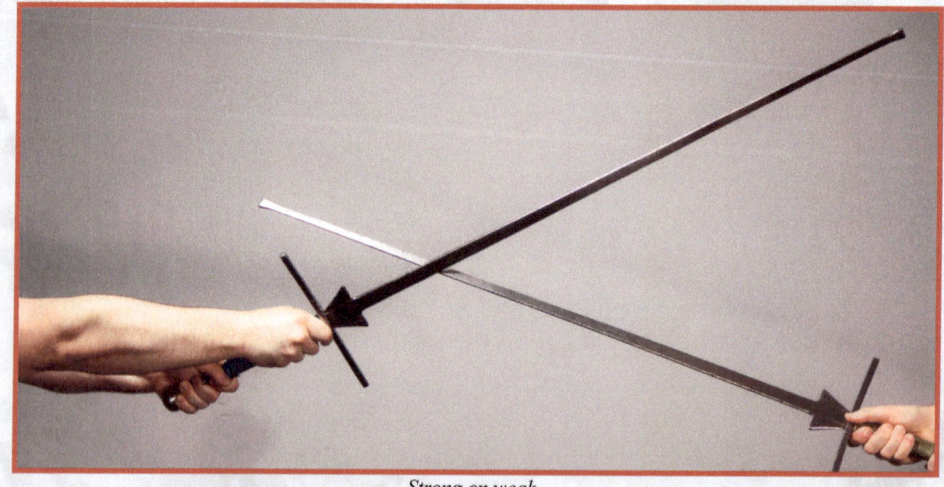

Strong on weak.

POSTURE

The masters do not elaborate in great detail regarding posture. Attempting to strictly define proper posture based solely on the illustrations found within the historical texts is difficult, as there doesn't appear to be any uniformity in how a fencer should stand. The advice offered here is based on commonalities observed in the way fencers are depicted in the historical sources.

A fencer should set one foot in front of the other, with a distance between them that is not so narrow that balance is compromised, and not so wide that stepping becomes difficult. There are illustrated examples of fencers adopting both forward and rear weighted postures, though it is impractical to lean in such a way that a fencer cannot easily change their posture or perform strikes. The back should remain straight and usually upright without hunching or "rounding" the shoulders forward.

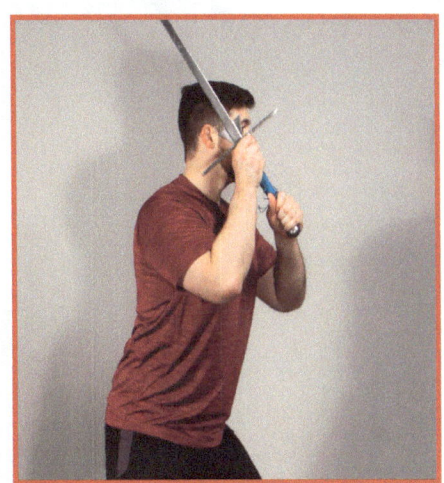

Stable knee position.

Collapsed knee.

Completed cut with control.

Posture comprised after cut.

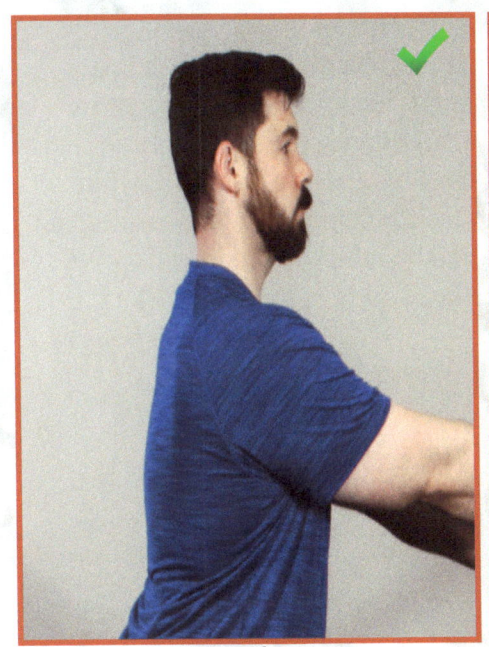

Proper upright posture.　　　　　　　　*Rounded back.*

FOOTWORK

Fencers should not idle in one place. Footwork should be in constant use to maneuver in the onset, enter the krieg, find openings in the krieg, and withdraw from the krieg when necessary. Meyer advises that all strokes should have their own step and "all combat happens vainly, no matter how artful it is, if the steps for it are not executed correctly." [1]

The *MS 3227a* warns against stepping directly towards the target when striking.[2] Fencers should instead step slightly to the side of their opponent. The manuscript also warns against stepping in such a manner in which the fencer cannot quickly recover and perform another strike. A fencer's footwork should be controlled and place them in the ideal measure for another attack.

Steps performed without an attack are not expressly detailed in the early glosses. Presumably, fencers would be constantly adjusting their position and posture with a series of both small and large steps before coming to blows. Joachim Meyer advises that there are steps which draw the fencer towards their opponent and steps with which the fencer withdraws from their opponent. Meyer also advises that a fencer should hide their intent with deceptive footwork.[3] For example, a fencer may act as if they intend to step deeply in one direction while in fact making a more shallow step or act as if they intend advance slowly towards their opponent before suddenly exploding towards them.

Passing Step

This step describes the rear foot moving past the lead foot, often in a somewhat lateral direction. When attacking from the onset, fencers are expected to begin with their non-dominant foot forward, and pass to their dominant side when executing a vorschlag. Ringeck writes that failure to follow an attack from zufechten without a passing step would result in an attack that is too short.[4] Passing steps can be used to withdraw from an opponent as well, pulling the lead foot back and behind the rear foot.

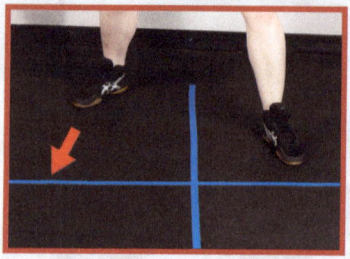 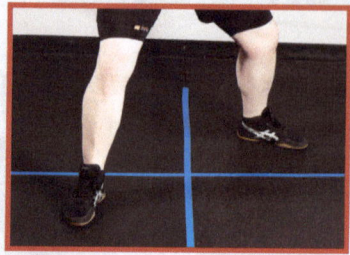

Right foot passes the left foot. The step carries the fencer slightly to their Right.

Left foot passes the right foot. The step carries the fencer slightly to their left.

Advancing Step

Some actions require a short step where the lead foot is advanced forward and somewhat laterally. Advancing steps can be used to power actions in situations where a passing step would carry the fencer too near their target. After the lead foot is advanced, the rear foot may also move forward if necessary to stabilize the fencer.

Left foot is advanced forward. The step carries the fencer slightly toward their left.

Advancing and passing steps have been defined by the author in an effort to describe simple footwork. While not explicitly defined in the glosses, these forms of footwork are commonly used to perform various actions.

TRIANGLE STEP

According to Meyer, the triangle step describes the rear foot moving behind the forward foot.[5] This step serves multiple functions but is perhaps most commonly used to turn the fencer in place following a particularly broad passing step. Not only can the triangle step allow a fencer to reorient themselves, it can also be used to turn the hips while performing an attack without having to adjust measure.

Fencer pulls their left foot behind their right, reorienting themselves with the triangle step.

Fencer performs triangle step after a broad passing step so they can reorient and stabilize themselves.

GRIPPING THE SWORD

Fencers should primarily squeeze with the pinky and ring finger so their grip resembles a handshake. Gripping the sword so tightly that it is wielded like a hammer is often detrimental to fencing. The *MS 3227a* warns against gripping the pommel and advises to keep both hands upon the sword's grip.[6] While keeping both hands on the grip is ideal when cutting, there are many illustrations depicting fencers gripping the pommel when thrusting. The dominant hand should be the one nearest the crossguard, though it doesn't have to be in contact with it. The fencers in the *Goliath Fechtbuch* are depicted gripping the sword with quite a bit of space between their dominant hand and the crossguard.

REVERSED GRIP / THUMB GRIP

Some actions appear to show the crossguard reoriented so that it is almost perpendicular to the wrist. The thumb may also be extended in these actions so that it is near the flat of the blade. There is no specific name in the glosses for this manner of grip, though Meyer does categorize cuts where the hands may be reoriented on the hilt as *reversed strikes.*[7] When extending the thumb while adopting a reversed grip, it is unwise to apply pressure with the pad of the thumb upon the blade. Attempting to support the blade with the thumb reduces the possible surface area of the palm upon the grip, robbing the grip of strength.

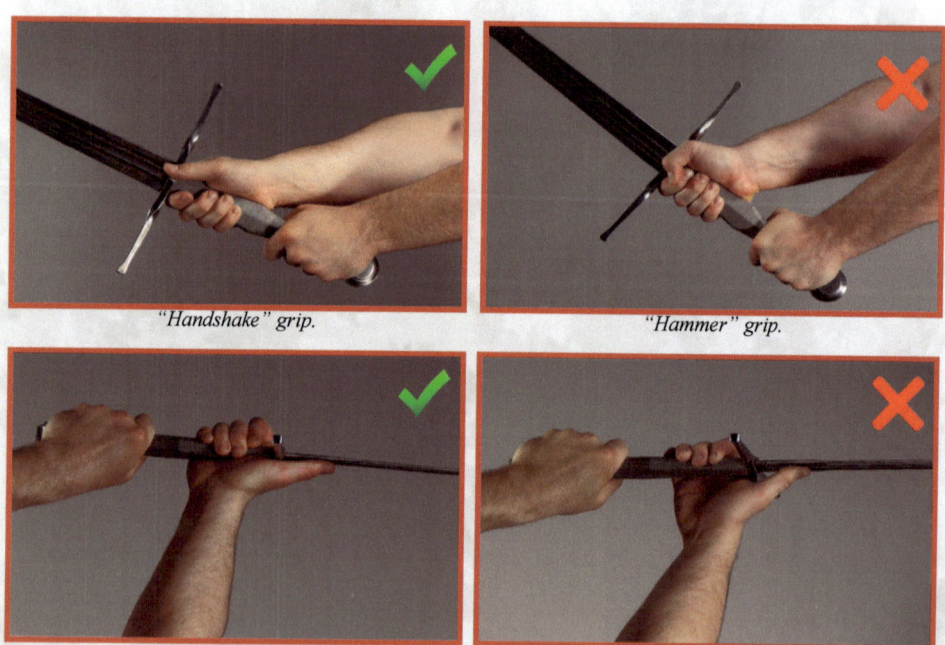

"Handshake" grip.

"Hammer" grip.

Reversed grip with thumb applying little pressure.

Improperly pressing the thumb into the flat of blade.

Four Openings

The openings are the target zones where a fencer directs their attacks. A vertical line down the center of a fencer divides the left and right openings. The upper openings are above the belt and the lower openings are below a fencer's belt. The four openings resulting from this division of the man are thus: upper-right opening, upper-left opening, lower-right opening, lower-left opening.

Where is the Belt?

When the belt or girdle is referenced in regards to the four openings, it is not easy to know exactly where this horizontal line is intended to be drawn on the human body. It is perhaps easiest to think of the belt as somewhere near the natural waist, where belts would have been worn during the time of this art. Upper opening targets typically include the head, neck, and chest. Lower opening targets are typically the lower torso and elbows. The legs are rarely targeted.

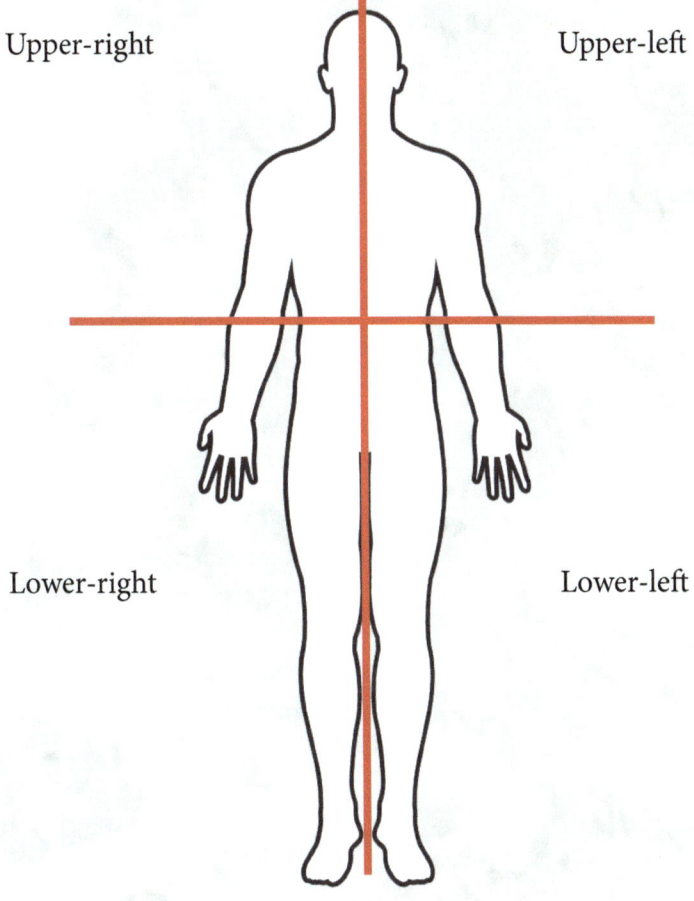

Upper-right

Upper-left

Lower-right

Lower-left

DREI WUNDER - THREE WOUNDERS

There are three primary methods of attacking: cutting, thrusting, and slicing. While fencing within the krieg, a fencer should not rush from one opening the other, but allow bind pressure to dictate which of the three wounders to direct to the appropriate opening. One attack leads to another and it is common that one type of wounder sets up an opportunity for a different wounder to another opening.

HAU - STRIKE/CUT

Strikes are attacks delivered in an arcing motion, which are intended to debilitate with severing or percussion. The *MS 3227a* warns against performing cuts while gripping the pommel as doing so "restrains the strike."[8] Pulling the pommel back while pushing the dominant hand forward creates a suboptimal cut and it is possible this passage from the *MS 3227a* is attempting to convey that.

Ringeck advises that "every thing which you wish to fence, conduct it with the entire strength of the body."[9] Instead of propelling the sword forward with a "push-pull" motion using the pommel as a lever, fencers should drive their cuts by turning the hips. More power is generated when large muscle groups, such as the core, are used to propel the sword, rather than relying on the arms alone.

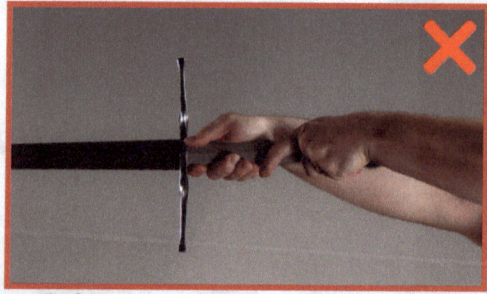

Broken wrist position as a result of push-pull cut.

A firm grip and unbroken wrist are important to secure the weapon and establish structure. The muscles in the arms are activated to create a stable structure. The *MS 3227a* advises that cuts move as though the edge is pulled by string into the opening[10], and as such, the hands should not lead in front of the blade where they can be struck before the attack is completed. If alignment of the edge is not maintained through the cut, the attack may fail to properly sever its target.

According to the *MS 3227a*, there are two primary cuts: The *oberhau* and the *unterhau*.[11] Cuts which descend are derived from the oberhau. Cuts which ascend are derived from the unterhau.

Oberhau - Overcut

When oberhau is used to describe a specific cut, rather than a generic term for descending cuts, it is typically performed with the long-edge and extended arms. The arc of a standard oberhau ends with the point near the ground.

Fencer begins with sword near shoulder. Grip is relaxed.

Fencer squeezes grip with bottom two fingers. Arms are extended forward with blade leading the hands.

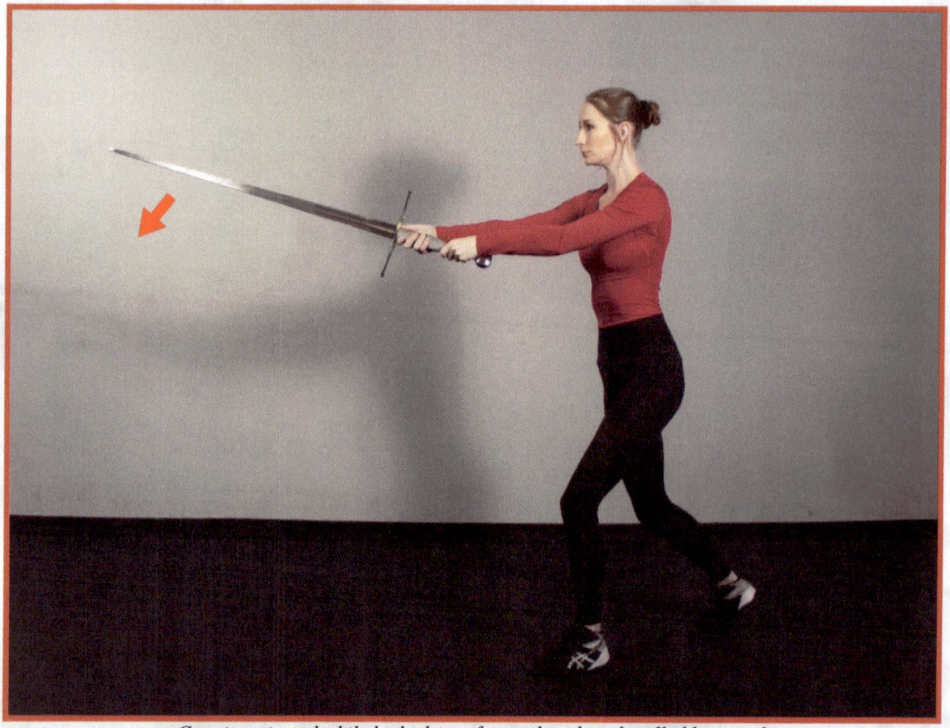

Core is activated while body drives forward as though pulled by sword.

Cut is completed with passing step.

UNTERHAU - UNDERCUT

When unterhau is used to describe a specific cut, rather than a generic term for ascending cuts, it is typically performed with the long-edge and extended arms. The arc of a standard unterhau ends with the point directed towards the sky.

Fencer begins with sword near shoulder.

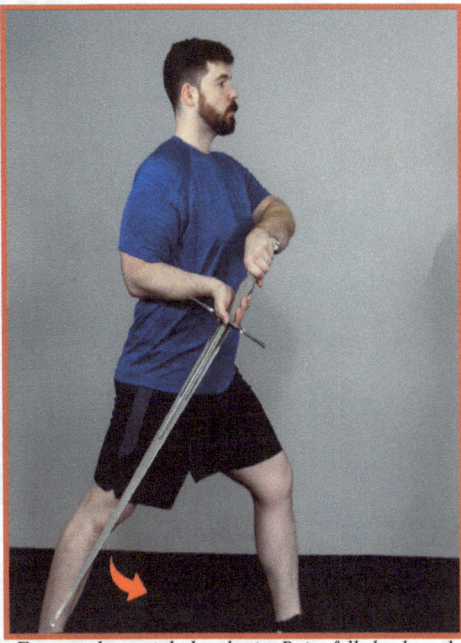

Fencer relaxes right hand grip. Point falls backward.

Fencer squeezes grip with bottom two fingers. Arms are extended forward with blade leading the hands.

Core is activated while body drives forward as though pulled by sword.

Cut is completed with passing step.

STICH - THRUST/STAB

The MS 3227a advises that thrusts move towards an opening as though the point was being pulled there by a string.[12] From the onset, a fencer extends their arms and allows the forward movement to pull their torso forward. While lunging over the lead leg, the fencer completes the thrust with a passing step which carries them forward and somewhat laterally.

Fencer begins with point extended forward and hands withdrawn.

Fencer extends point and then leans slightly forward.

Fencer completes thrust with passing step.

EINSCHIESSEN - SHOOTING IN

Thrusts are often set up using a cut. If a fencer performs an oberhau and their opponent binds upon it, the attacker can "shoot-in" with the point if they feel that their opponent is soft in the bind. Striking with power allows the attacker to place their point in an ideal position for a thrust, potentially even displacing the opponent's sword.

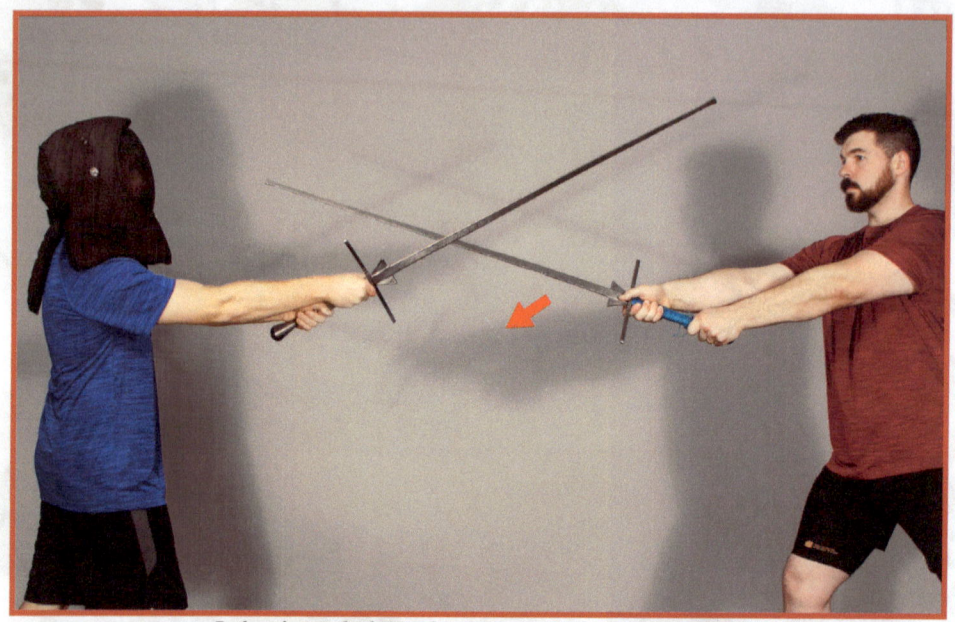

Red performs oberhau with passing step and is parried by Blue

Red advances right foot while maintaining structure and drives point forward.

SCHNITT - SLICE

Slices are not draw cuts; They are attacks used to control an opponent. Pressing with the edge, a fencer can manipulate an opponent's arms with the slice and create opportunities to follow up with other attacks. Proper slicing begins with structure in the arms and is completed by driving with the body.

ZETTEL

Young knight learn
To have love for God, honor women
So your honor grows
Practice chivalry and learn
Art which adorns you
And in war curries favor to honor
Wrestle well grappler
Lance, spear, sword, and knife
Wield valiantly
And make useless in others' hands
Hew in and hurry there
Rush in, land a hit, or let go
So that the one maturing in wisdom
Sees praise
Upon his grasp
All arts have distance and measure

If you want to show art
Go left and right with hewing
And left with right
If you wish to fence strongly
Whoever goes after hews
He permits his art little joy
Hew nearing what you want
No change comes into your shield
To head, to body
Do nor abandon the light hits
With entire body
Fence what you wish to perform strongly
Hear what is simple
Do fence above left if you are right
And if you are left
Also very clumsy in the right

Before and after, the two things
Are a wellspring of all art
Weakness and strength
INDES, with it note that word
So you may learn
To work and assess with art
If you frighten easily
Never learn any fencing
Learn the five hews
From the right hand, whoever wards it
Him we praise
Wishing to be worthwhile in the art

Zornhau, Krumphau, Zwerchau
Has Schiller with Scheitler
Fool parries
Nachreisen, Uberlauffen set hew
Durchwescheln, Zucken
Durchlauffen, Abschneiden, Hend Trucken
Hang, Wind with the openings
Strike, catch, sweep, stab, with shoving

Whoever over hews you
ZORNHAU threatens him
If he becomes aware of it
Take it off above without danger
Be stronger against
Wind, stab, if he sees it, take it below
Note this precisely
Hew, stab, position, soft or hard
INDES, and move after
Without hurry, your war be not rushed
Whoever aims the war above
He will be shamed below
In all windings
Learn to find hew, stab, slice

You shall also with
Testing, hew, stab, or slice
In all meetings
If you want to fool the masters

Know four openings
Aim so you strike wisely
Without any danger
Without regard for how he bares

If you want to exact revenge
The four openings artfully break
Duplicate above
Correctly transmute below
I say to you truthfully
No one protects themselves without danger
If you have heard
He may rarely come to strike

KRUMP on nimbly
Throw the point upon the hands
Krump whoever sets well
With stepping, it destroys many hews
Krump to the flats
To the masers, if you want to weaken them
When it clashes above
Stand off, that I wish to praise
Not Krump, short hew
Show disengage with it
Krump whoever strays you
The noble war confuses him
So that he truthfully
Does not know where is without danger

ZWERCH takes
What comes from the day

Zwerch with strength
With it note your work
Zwerch to the plow
Join hard to the ox
Whatever crosses itself well
Threatens the head with jumping
Failer leads the weapon well
It touches from below as planned
Inverter forces
Run through, also wrestle with
The elbow
Take wisely, jump to him in balance
Failer twofold
If one connects, make with slice
Twofold it continues
Step in left and do not be lazy

SCHILLER breaks in
Whatever buffalo strike or stabs
Whoever threatens change
Schiller robs him of it
Schiller, if he shortens to you
Disengage defeats him
Schiller to the point
And take the neck without apprehension
Schiller to the upper
Head, if you want to destroy the hands

The SCHEITLER
Is dangerous to the face
With its turn
Quite dangerous to the chest
Whatever comes from him
The crown takes it away
Slice through the crown
So you already counter it hard

Press the sweeps
Withdraw them with slicing

VIER LEGER alone
Hold from then and flee the common
Ox, plow, fool
From the day, are not unknown to you

Four are VERSTZEN
Which also greatly injure the positions
Beware of parries
If it happens, it hurts you more
If you have been parried
And how that has happened
Hear what I advise
Wrench off, hew quick with threat
Plant four ends
Remain upon them, learn, if you to end

Learn to NACHREISEN
Twofold, or slice into the defenses
Two external figures outer takings
Thereafter the work begins
And test the movements
Whether they are soft or hard
Learn the feeling
INDES, that word slices sharp
Nachreisen twofold
If one connects, make with the old slice

Whoever aims below
UBERLAUFFEN, then he will be shamed
When it clashes above
Strengthen, that I indeed praise
Make your work
Or press hard twofold

*Learn to **ABSETZEN***
To artfully destroy hew, stab
Whoever stabs to you
Your point lands and his breaks
From both sides
Land a hit every time you want to step

*Learn to **DURCHWECHSELN***
From both sides, stab with severity
Whoever binds onto you
Durchwechseln nearly finds him

Step near in binds
***ZUCKEN** gives good finds*
Retract, if he connects, retract more
Work, devise, that does him woe
Retract in all meetings
To the masters, if you want to fool them

***DURCHLAUFFEN**, let hang*
Grim with the pommel if you want to wrestle
Whoever strengthens against
Durchlauffen, with it note

ABSCHNEIDEN
Slice off the hard
From below in both movements
Four are the slice
With two below, two above

HEND TRUCKEN
Turn your slice
Press the hands to the flats

*Two **HENGEN** will be*
Out of one hand

In all movements
Hew, stab, position, soft or hard

Make the speaking window
Stand confidently, inspect his play
Strike him so that it snaps
Whoever withdraws themselves in front of you
I say to you truthfully
No on protects themselves without danger
If you have heard
He may rarely come to strike

Whoever leads well and correctly counters
And finally indeed directs
And counters individually
Each in three wounders
Whoever correctly hangs well
And with it brings **WINDEN**
And eight are the Winden
Consider with correct scales
And to them one
I mean, triple the same Winden
So they are twenty
And four, in their own count
From both sides
Learn eight Winden with steps
And test the movements
Not more than soft or hard

-Translation by Stephen Cheney

FIVE STRIKES

The first five haupstucke presented in the Zettel are the five strikes which Meyer refers to as the meisterhau, or *master strikes*. These are considered secret attacks which Danzig and Ringeck advise few masters know of. When executed from the onset, the five strikes are typically performed from the fencer's dominant shoulder.

ZORNHAU - WRATH CUT
The zornhau is a descending true-edge strike that counters other oberhau. Lew advises this is not a parry and the focus of this cut is to present your point in a threatening manner. The early glosses do not describe how to step with this strike and it is possible the attack may be executed with a passing step or small lateral step with the front or back foot.[13]

Fencer begins with sword near right shoulder.

Cut descends in a very linear angle.

ZORNHAU TRAJECTORY
The zornhau is perhaps most effective when delivered in a vertical line through target's center with very little angle. Attacking in this manner ensures the fencer is targeting their opponent (rather than the sword) and maintains a threatening point. The fencer should not prematurely terminate the cut's trajectory until pressure in the bind dictates the next action.

Zornort - Wrath Point

If a fencer counters an oberhau with the zornhau and feels the opponent is soft in the bind, then they can attack with a thrust delivered straight forward to the opponent's face or chest. Assuming the zornhau was delivered with a small lateral step, the zornort is then executed with a passing step which drives the point forward with explosive speed.

Blue performs Zornhau with small lateral step.

Blue feels Red is soft in bind and drives zornort straight forward with passing step.

WINDING THE POINT

If the target of the zornhau remains in the bind with strength, rather than attempting to thrust the point away, the one performing the zornhau should raise their hands in the bind.[14] Raising the hands in this manner allows the fencer to place the strong of their sword against the weak of their opponent's blade. From this position, the fencer should attempt to stab their target in the face.

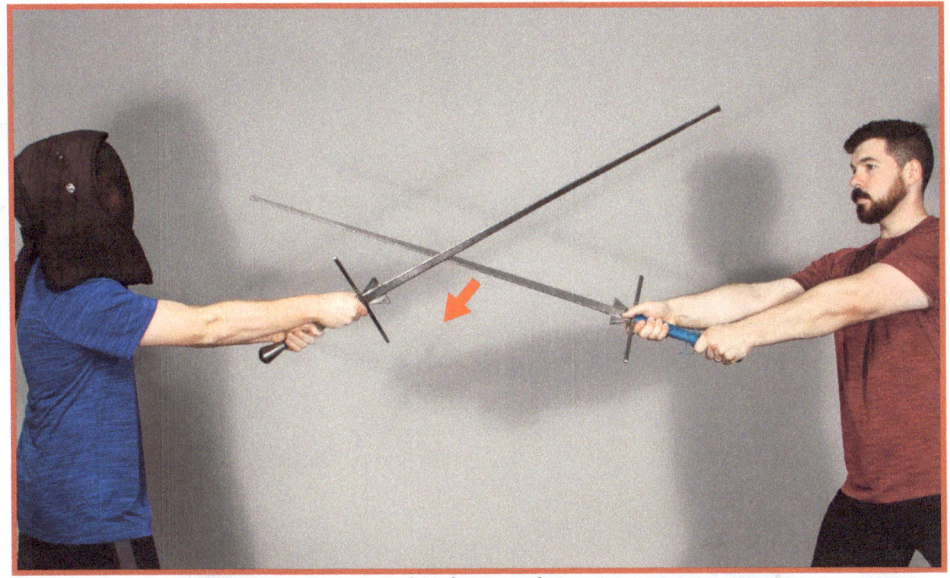

Red performs zornhau.

Blue resists in bind. Red raises hands and performs zornort.

Abnehmen - Taking Off

If the zornhau or zornort is defended with a hard lateral pressure, the attacker can lift their sword up and remove it from the bind. The fencer completes the abnehmen action with a cut, which descends to the other side of the opponent's sword and is directed to the head. Stepping laterally can assist in powering this cut and positioning the fencer in the ideal measure to deliver it.

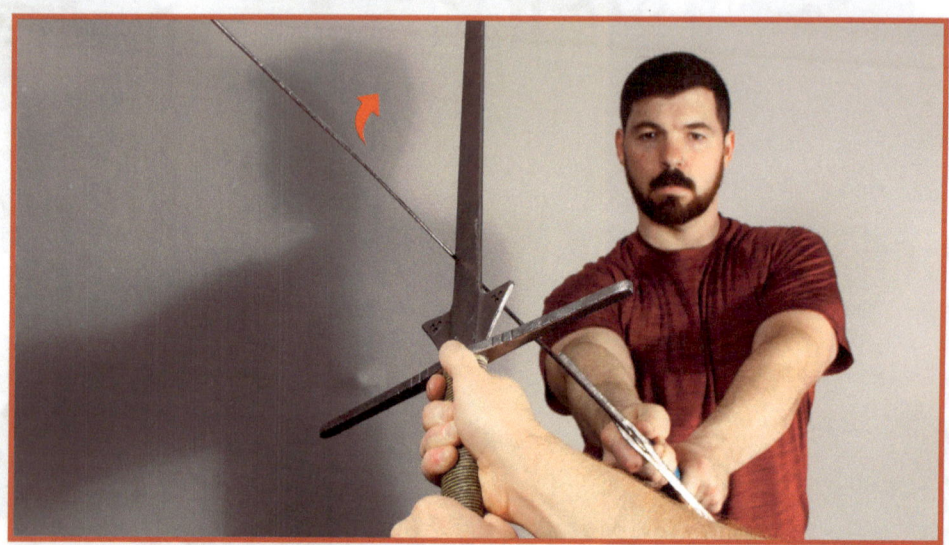

Red drives the zornort forward and Blue sets aside with firm lateral pressure.

Red lifts off bind while stepping to their left and cuts around the bind.

KRUMPHAU - CROOKED CUT

When executed from the fencer's dominant side, the krumphau is a long-edge strike performed with crossed arms. The krumphau is used in conjunction with lateral footwork, which propels the fencer away from attacks. One application of the krumphau is to strike the opponent's hands as they attempt an unterhau or oberhau.

Fencers stand in vom tag.

Blue performs an oberhau while the Red springs laterally and cuts Blue's hand with a krumphau.

SCHRANKHUT - BARRIER GUARD

The krumphau can be performed from both the shoulder and from a guard called *schrankhut*. When standing with the non-dominant foot forward, the fencer holds their sword towards their dominant side with the point on the ground and the long-edge facing up. When standing with the dominant foot forward, the sword is held to the fencer's non-dominant side with the false-edge facing up.

Fencers standing in schrankhut invite attacks to their upper-openings which can be displaced with the krumphau. When performed from schrankhut on the fencer's non-dominant side, the krumphau is a false-edge strike with uncrossed arms. The trajectory of the krumphau is such that it ends in schrankhut to both sides.

STRIKE TO THE FLATS

When threatened with an oberhau, a fencer can deliver a krumphau into the flat of an incoming strike. From the ensuing bind, the one performing the krumphau can unwind the pommel from under their wrist and deliver a false-edge cut to their opponent's left openings. A fencer can also wind their false edge upon the opponent's sword and stab the chest.

THRUST OR CUT?

Deciding to perform a cut or thrust following a krumphau should be decided by measure. A fencer may be too close to thrust or too far away to cut after binding upon the opponent's sword.

Blue deflects and suppresses oberhau with krumphau.

Blue pulls left foot behind right foot with triangle step while turning body and cutting with false-edge.

Krumphau deflects and suppresses oberhau.

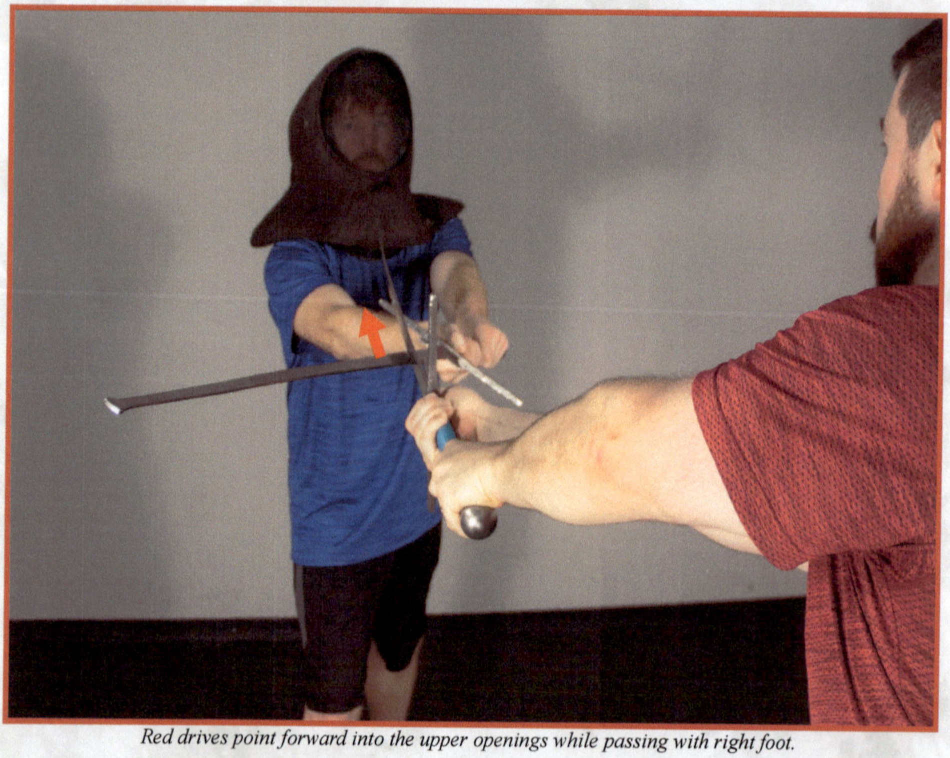

Red drives point forward into the upper openings while passing with right foot.

ZWERCHAU - THWART CUT

This attack is capable of countering descending cuts. When attacked by a descending cut, the fencer steps toward their dominant side and delivers a strike with the false-edge. The crossguard is turned in front of the fencer's head in such a manner that the thumb is shielded beneath the blade. With hands held high, the fencer catches the threatening strike near their hilt and cuts their target upon the head. When performing the zwerchau, it is important to keep the pommel higher than the point.

Fencer begins in vom tag.

Fencer pulls hands across chest.

Fencer extends sword towards target.

Fencer completes cut with passing step.

THWART TO OX AND PLOW

The zwerchau can be delivered to all four openings. Striking to the ox describes cutting to the upper-openings while striking to the plow is when the zwerchau targets the lower openings. Because the zwerchau is performed without fully extending the arms, it is an excellent nachslag while winding in the krieg. When performed with lateral stepping, the zwerchau can safely be performed in a continuous series of blows with each strike targeting an alternate opening. The zwerchau works best when delivered from one side to the other and moving back and forth from the upper and lower openings.

ZWERCHAU ANGLES

While the zwerchau can be performed along a horizontal line, it will typically offer more utility when descending or ascending into an opening. Descending zwerchaus performed with the hilt held high allows the fencer to drive the strong of their sword into the weak of their opponent's blade. Achieving the same bind with a horizontal strike would result in the cut harmlessly passing above the target. Attempting to strike low with a horizontal cut leaves the fencer's upper openings exposed, while an ascending zwerchau allows the attacker to keep their hilt high so their head is protected. Horizontal zwerchaus are still useful when delivered to the upper openings without having to intercept an opposing strike.

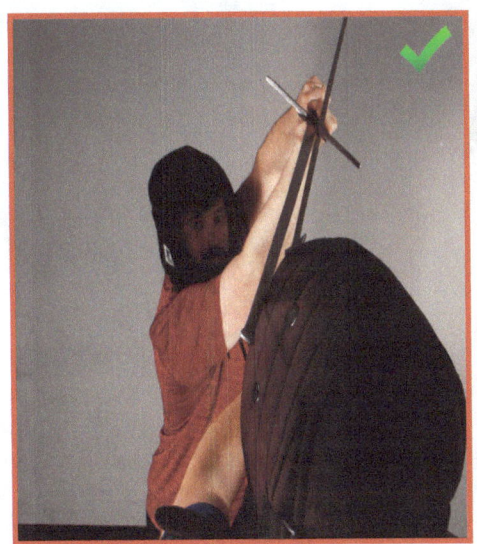

Descending zwerch with high hands.

Horizontal zwerch with high hands.

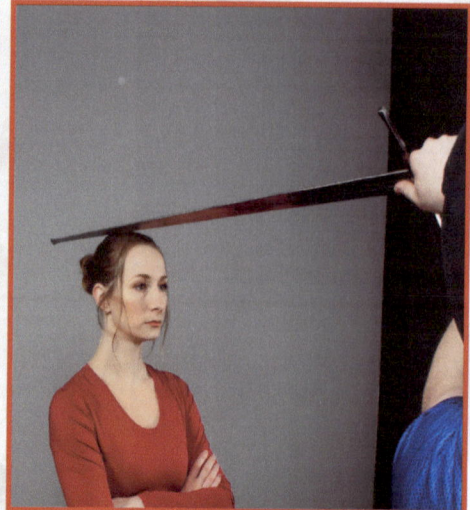

Descending zwerchs.

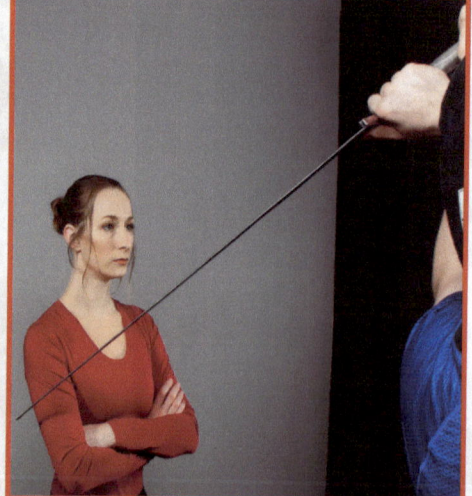

Ascending zwerchs.

FELER (FAILER)

The feler is a feigned strike that sets up another strike. One example of this is when a fencer provokes with a zwerch to his opponent's upper-left opening, and pulls it short before performing another zwerch to the upper-right opening. Another example of a feler is when a fencer provokes with a descending cut to their opponent's upper-left opening. Before the oberhau can be parried, the fencer pulls the cut back and performs an ascending zwerch to either lower opening, or strikes the opposite upper opening with another oberhau.

EFFICIENT FEINTING

Often times fencers will fail to perform feints correctly by either stepping too near their opponent while provoking or by half-heartedly presenting the feint in such a manner that the opponent can obviously tell it is a feint. Fencers should be near enough to their opponent that the feint is threatening, but avoid stepping into the krieg until they are certain the opponent has been provoked into parrying. Assuming the opponent will parry can be dangerous as the defender may counter the feint with a committed attack. The fencer performing the feint is most vulnerable while transitioning between the provoking action and a commited followup attack.

Red performs a zwerchau as a feint.

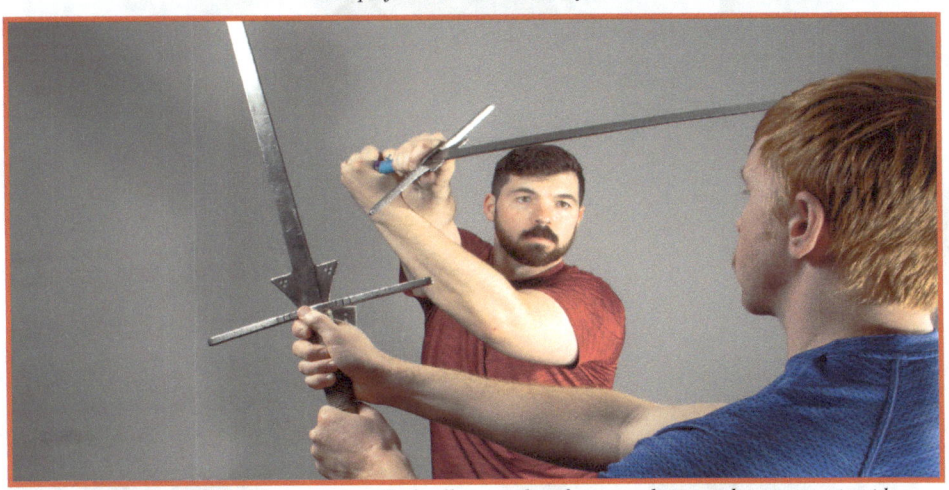

Blue attempts to defend first zwerchau. Red pulls cut and performs another zwerchau to opposite side.

Fencers begin in vom tag.

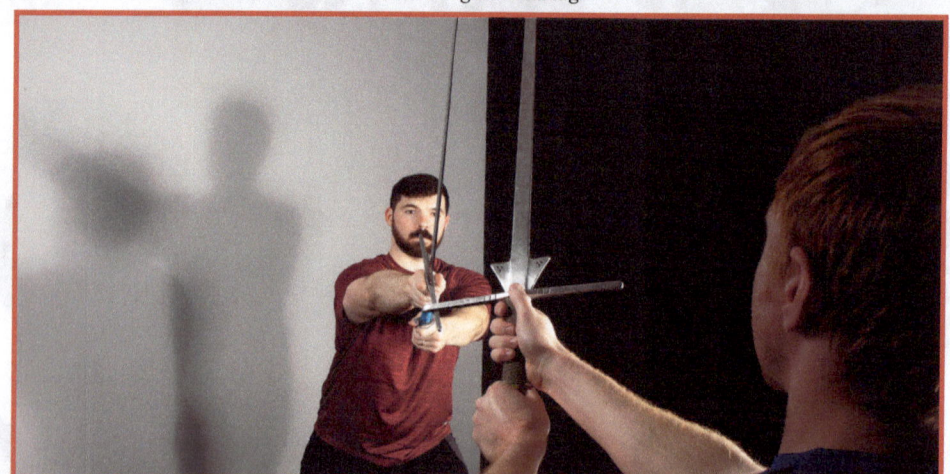

Red feints with an oberhau performed while advancing left foot.

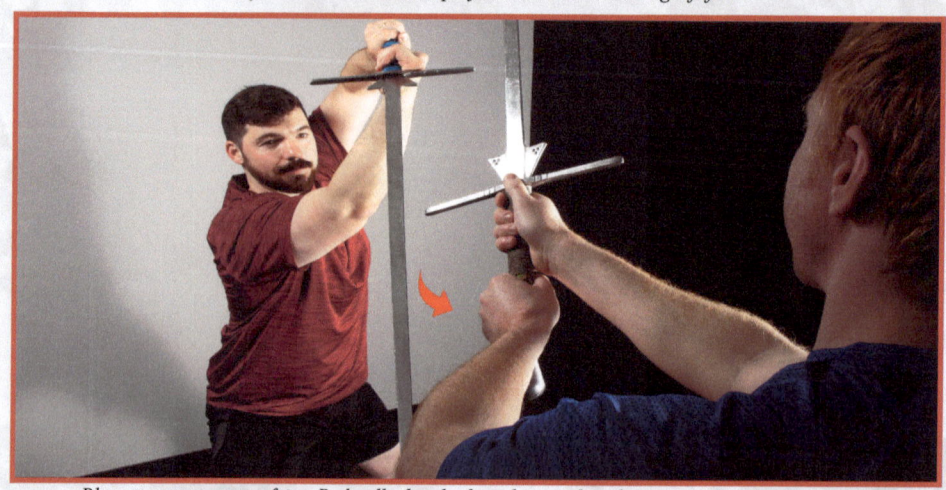

Blue attempts to parry feint. Red pulls the oberhau short and performs zwerch to lower opening.

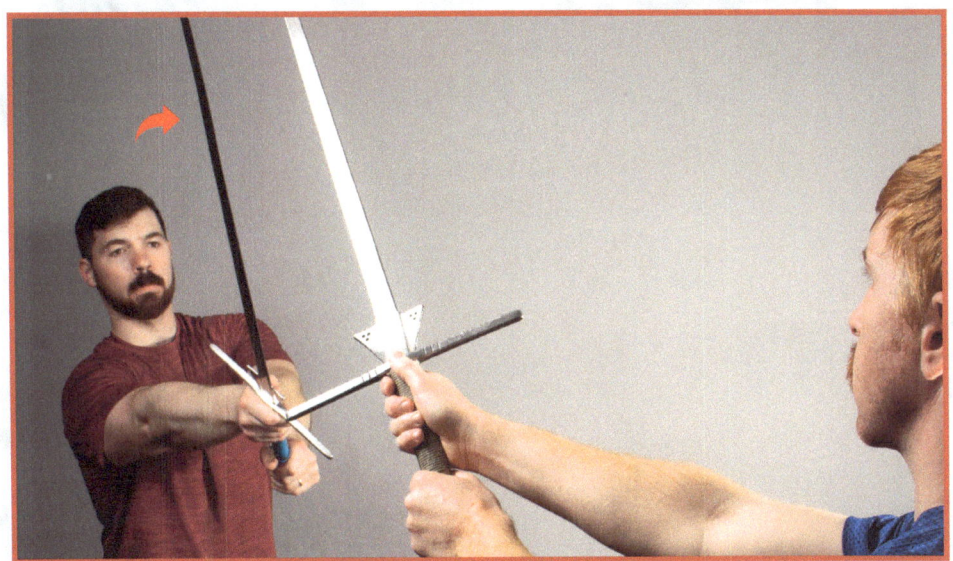

Red feints with oberhau. Blue attempts to parry.

Red pulls oberhau and performs another oberhau to opposite side.

COUNTER-ZWERCH

When a fencer is preparing to perform an oberhau and sees their attack will be met with a zwerchau, they can fall upon the zwerchau with their long-edge. If their opponent performs another zwerchau to their upper-right opening, they should counter this with a zwerchau of their own. This zwerchau should be delivered to the left side of the opponent's neck with the false-edge, under the opponent's sword.

DESCENDING COUNTER

If a fencer attempts to counter a zwerchau with a horizontal zwerchau, both fencers will likely strike one another as a result. When the counter-zwerch descends into the neck with the hands held high, it is much easier to create a barrier with which a fencer can protect themselves.

Descending zwerchau successfully defends against opponent's attack.

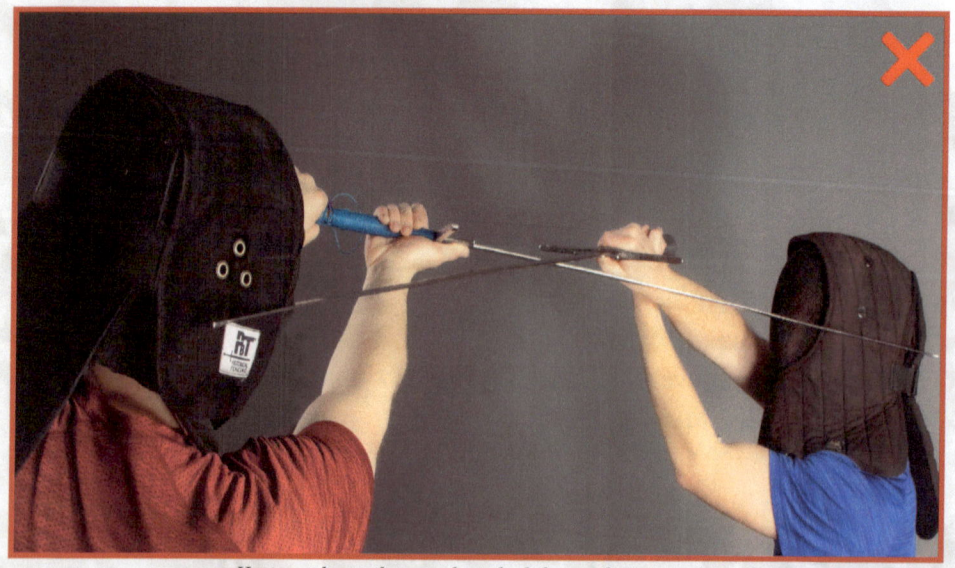

Horizontal zwerchau results in both fencers being struck.

Schielhau - Squint Cut

This schielhau, or schiller, is a descending false-edge strike which Ringeck advises is a good counter to the 'buffalo': a fencer who overcomes others with violence and power. Similar to the zornhau, footwork for the schielhau is not always described in the glosses and it is possible to perform the cut with the non-dominant foot remaining forward.[15] Unlike the zwerchau, the schielhau is a cut delivered with extended arms.

Oberhau Counter

When an opponent attacks with a descending cut from their dominant side, a fencer can counter with a schielhau directed towards the opponent's head or right shoulder. Ringeck describes striking into the weak of the opponent's blade, while Danzig and Lew advise to cut over the opponent's sword. These RDL descriptions of the schielhau suggest that the hands are somewhat elevated while remaining extended as the two swords meet.

Thrust Counter

Because the schielhau is performed with extended arms, it can be used to punish those fencing with retracted arms in certain situations. One example of this is when a fencer is performing a schielhau against a descending strike, and the opponent pulls the oberhau short or disengages from the bind with the intention of thrusting to the other side. The fencer performing the schielhau is already extended with their arms, allowing them to thrust into the chest of their opponent before they can fully disengage and attempt a thrust of their own.

The schielhau can also be used to counter an opponent attempting to directly thrust into them. If a fencer observes that their opponent is drawing their sword back in preparation for a thrust, they should immediately execute the schielhau and stab into the chest or face before their opponent can complete the thrust. If an opponent is already extending their arms with the point threatening, a fencer can schielhau onto the opponent's hands (or sword) and stab into the throat with a passing step.

SCHEITELHAU - PARTING CUT

The scheitelhau, or scheitler, is a descending long-edge strike delivered to the opponent's head when they are standing in such a manner that their sword is low with the point near the ground. This cut is performed with a passing step towards the opponent. The defender may lift their crossguard high in the air to defend the scheitelhau using a posture knows as *kron,* or crown. Once countered with kron, the attacker may lift their pommel high and stab downward towards the head or chest. If the defender continues to raise their sword, the attacker can dip their point beneath the opponent's hands and thrust.

SCHEITLER MECHANICS

There is some disagreement among modern practitioners regarding how the scheitelhau should be performed.[16] If the cut threatens the head and forces the opponent to parry, setting up the following atacks, the mechanics are probably sound.

Blue executes Scheitelhau. Red intercepts with kron.

Blue lifts pommel upward with left hand and plunges point towards Red's chest.

Red lifts higher with hands to defend against thrust. Blue disengages from bind and thrusts under Red's hands.

49

Vier Leger - Four Positions

There are four primary positions or guards. Fencers are expected to fence solely from these positions. Meyer advises the *vier leger* are the chief postures from which all other fencing positions derive.

It perhaps seems unintuitive to describe the vier leger after teaching the five strikes and their plays. One could infer from the way in which the Zettel orders the haupstucke, that it is more important to understand how to attack than it is to define positions. Or, the order of the Zettel is completely arbitrary, though that seems unlikely. What should be stressed however, is that it is unproductive to obsess over the vier leger. There is no uniformity in their depiction and attempting to adopt each position "perfectly" is a fool's errand.

Ochs - Ox

There are two variations of this position. When standing with the non-dominant foot forward, a fencer holds their sword to their dominant side with the crossguard in front of the head. The hands are held high enough so that the point can hang towards the opponent's face. This position can also be similarly assumed with the dominant foot forward and the sword held to fencer's non-dominant side.

PFLUGH - PLOW

Similarly to ochs, there are two variations of this position. When standing with the non-dominant foot forward, a fencer holds their sword on their dominant side, near the hip. The point of the sword is held upright, threatening the face of the opponent. While held on the dominant side with crossed arms, the false-edge is facing up. When this position is assumed on the other side, the true-edge faces up.

ALBER - FOOL

This guard is assumed by standing with the dominant foot forward. The arms are outstretched and the point rests near the ground.

VOM TAG - FROM THE DAY

This posture is where all of the five strikes can be performed. A fencer adopts it by placing their non-dominant foot forward and holding their sword near the dominant shoulder or above the head.

Vom Tag Variations

VIER VERSETZEN – FOUR DISPLACEMENTS

These are the four cuts which are intended to counter the vier leger. These cuts have been detailed earlier and their relation to the vier leger are thus:

- Krumphau counters ochs.
- Zwerchau counters vom tag.
- Schielhau counters plugh.
- Scheitelhau counters alber.

TIMING THE VERSETZEN

One possible way of considering the vier versetzen, is that they are used as a vorschlag and performed while someone is passively standing in a specific position. Another possibility is that the vier versetzen are cuts performed immediately before a fencer commits an attack. This is supported by the Zettel and its glosses — which warn against parrying — while discussing the vier versetzen. Danzig specifically warns against parrying with the point high or to the side in a non-threatening manner. Ringeck writes in his gloss concerning the vier versetzen, that a fencer should perform cuts against cuts and thrusts against thrusts.

Using the vier versetzen to attack someone passively waiting in a guard carries potential risk. A zwerchau delivered at a fencer in vom tag that is not preparing to commit an oberhau is vulnerable to having his hands struck with the krumphau or a similar cut. Fencers laying in alber may potentially be waiting upon the scheitelhau in an effort to sweep it away.

If the vier versetzen are not attacks intended for targets idly waiting in a guard, one must then consider what manner of actions are likely to derive from the vier leger or end in the vier leger. and how the vier versetzen counter them. Rising cuts ending in ochs or a similar position can be countered with a krumphau, according to Ringeck. The zwerchau is deliberately intended to counter the sort of descending cuts a fencer would typically execute from vom tag. Those chambering thrusts from plugh can be countered with a thrust from schielhau, according to the glosses. And finally, a descending cut which ends short of its target and terminates in alber or a similar position can expose the attacker to a scheitelhau.

NACHREISSEN - RACING AFTER

Nachreissen is a concept which explains how a fencer can seek openings as their opponent presents them. Taking advantage of opportunities created by the opponent's movement is an effective way to attack, and carries less risk than striking at an opponent which is already poised to respond. Many of the examples of nachreissen provided in the glosses appear to be closely related to the vier versetzen.

GUARD CHANGE
When an opponent is positioning themselves to prepare an attack, a fencer can commit an attack of their own while their opponent is still attempting to enter the appropriate posture. A fencer can safely attack the upper openings with a cut or thrust while their opponent is in the process of lifting their sword to prepare a cut. If the opponent is rushing forward while lifting the sword high, a fencer may also fall upon their wrists with the slice before they can come down with the cut. An opponent drawing their sword back in preparation of a thrust can be safely stabbed before they can complete the attack.

FAILED CUT
An opponent who misses with their cut due to their inability to reach their target is vulnerable to counter-attack. These cuts may miss as a result of the target moving away from the attack or because the attacker misjudged measure. Regardless, a fencer should quickly strike their opponent's upper openings before they can draw their sword back up.

Uberlauffen - Overrunning

Uberlauffen governs how a fencer should strike from zufechten. Attacks made to the upper openings have greater reach than those made to the lower openings. It is therefore important to fence to the upper openings before closing into the krieg.

When an opponent cuts or thrusts to the lower openings from zufecthen, a fencer should not seek to parry or otherwise bind upon their sword, but instead cleave the attacker in the head with a powerful cut. This counter-cut should reach the opponent's upper openings before the opponent can reach the lower openings. A cut delivered to the upper openings with proper force should seriously compromise the quality of the opponent's attack and may potentially end the threat altogether.

Ansetzen - Setting the Point

A fencer can also thrust into the upper openings as a counter to low attacks delivered in zufechten. Setting the point upon the upper openings with extended arms is also an effective counter to those attempting to disengage under a fencer's a sword.

Absetzen - Setting Off

Absetzen refers to parries which allow a fencer to seize the vor away from their opponent in the moment the attack is offset. A fencer performing absetzen should maintain a point-forward position throughout the action so that they can quickly threaten with a thrust before the opponent can deliver a nachslag.

Setting Off the Thrust

When an opponent is preparing to thrust, a fencer may stand against them in pflug with the non-dominant foot forward. As the opponent extends forward with the thrust, the fencer standing in pflug can wind their sword across their body and offset the thrust. Passing forward with the dominant foot while setting off the thrust, the defender can perform a thrust of their own before the attacker attempts a nachslag. Moving somewhat laterally with the passing step can further protect the defender.

Red is threatened with thrust. *Red winds sword across body and steps with pass.*

Setting Off the Cut

If an opponent cuts into a fencer standing in pflugh to either side, the defender can perform absetzen by winding their sword upward against the strike. As the defender winds laterally into the incoming cut, they should drive high with their hands while keeping the crossguard past their head. If the defender must wind their sword across their body while performing the absetzen, they complete the setting-off with a passing step forward and thrust into the attacker. If the defender does not have to wind across their body, they thrust forward while advancing the lead foot.

Blue stands in vom tag. Red stands in pflugh.

Red counters oberhau with absetzen while performing an advancing step towards their left side.

Durchwechseln-ChangeThrough

Durchwechseln describes the point traveling under the opponent's sword and threatening to the other side. This is often the result of a fencer disengaging from a bind so that they can free their point and thrust into the other side of the opponent's blade. Disengaging from the bind in this manner is dangerous if the opponent is actively attempting to wind their point into openings. Durchwechseln works best when the opponent is pressing strongly in a lateral direction while moving their point in a non-threatening direction. It is possible to durchwechseln multiple times against an opponent that chases after the point and never seeks the vor themselves.

Blue presses Red's sword away.

Red disengages beneath Blue's sword.

Durchwechseln Measure

Disengaging with the point works best when the distance between the fencer and their target does not require them to withdraw their hands all the way back toward themselves. In fact, durchwechseln is faster and more efficient when performed with minimal motion. The fencer should keep the arms somewhat extended and rely on the wrists to move the point, rather than trying to pull the hands completely back.

ZUCKEN - PULLING

While durchwechseln describes disengaging under a bind, *zucken* is when a fencer disengages *over* a bind. An example of zucken is when a fencer extends their arms forward with a cut or thrust, and realizes that their target wishes to counter the attack with a strong parry that does not threaten with the point. The attacker can deny the bind by pulling their hands back, allowing the defender sweep their sword laterally without contact. The attacker can drop the point on the other side of their opponent's sword and extend their arms with a thrust.

Fencers begin in vom tag.

Blue threatens with oberhau.

Red attempts to parry attack. Blue pulls sword back to deny bind.

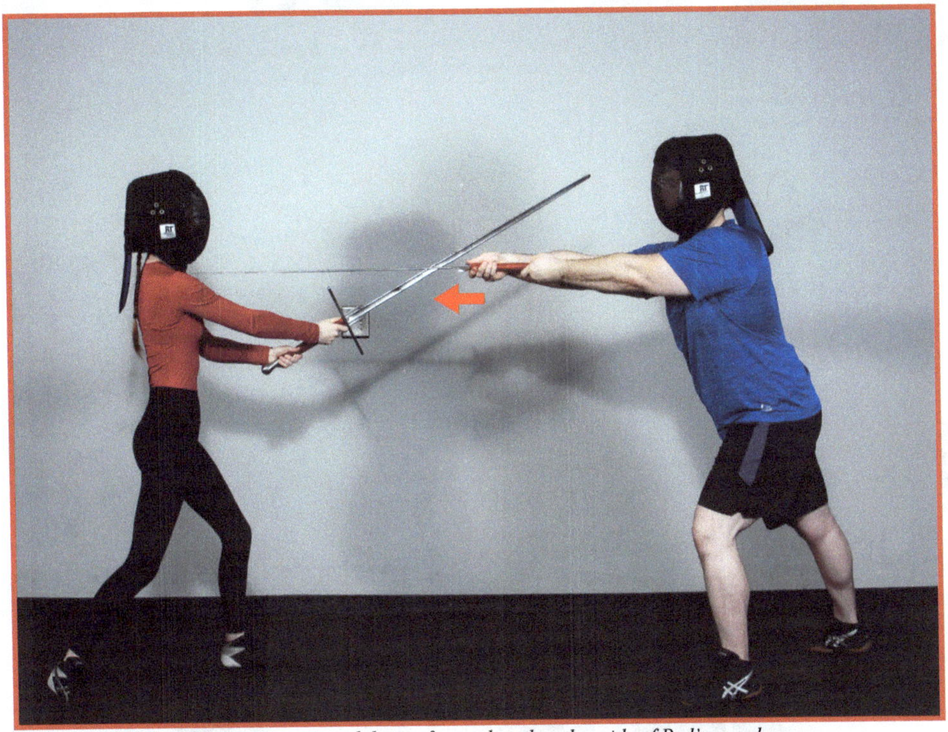

Blue drops point and thrusts forward on the other side of Red's sword.

CUTTING FROM ZUCKEN

If a fencer cuts with extended arms and is parried, they can pull their hands back and allow their sword to disengage above the bind. Once the sword has left the bind, the fencer performing zucken can cut downward with the long-edge into their opponent's opposite upper opening.

Fencers bind upon one another with extended arms.

Blue pulls hands back, allowing sword to disengage over the bind.

Blue cuts around to the other side of the bind with passing step.

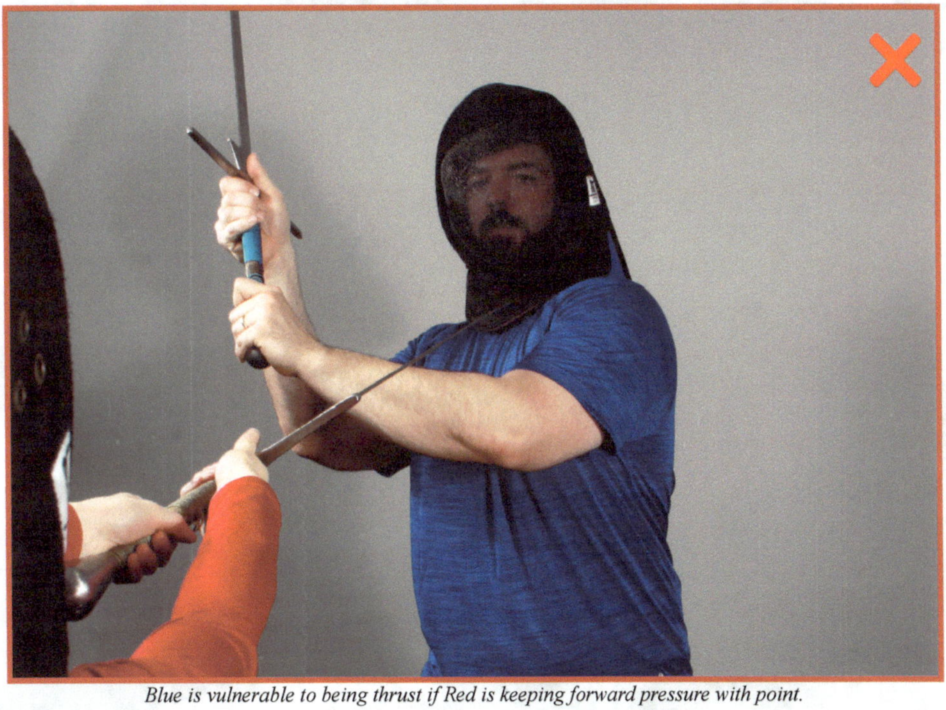

Blue is vulnerable to being thrust if Red is keeping forward pressure with point.

DECEIVING THE MASTERS

Zucken is described as a useful technique against master fencers who patiently wait in the bind and seek openings when their opponent leaves the bind. When a fencer moves into the krieg, they can partially pull their sword back without leaving the bind. While the opponent waits to race-after with a thrust or slice, the fencer performing zucken raises their hands and thrusts forward.

Meyer describes a similar action which uses zucken to deceive an opponent who will chase cuts. After binding on the opponent with a cut, a fencer can feign a strike to the other upper opening. Before this cut can be parried, the attacker terminates the attack and cuts back into the original upper opening.

Blue performs zucken, pulling sword partially back without fully leaving bind.

Red waits for Blue to leave bind. Blue instead thrusts back into original opening.

DURCHLAUFFEN - RUNNING THROUGH

Ringen (wrestling) is often necessary when a fencer rushes forward with the intention of overpowering their opponent. One example of how to counter someone rushing near with their hands held high is by releasing the dominant hand from the sword and allowing the blade to fall behind the back. The defender must keep their non-dominant hand high while allowing their blade to hang, thus creating a barrier against their opponent who is poised to strike from above.

  After allowing the sword to hang, the fencer performs *durchlauffen* by stepping forward with their dominant foot, past the dominant foot of their opponent. A fencer running-through in this manner must lower their center-of-gravity so their head passes beneath the dominant arm of the opponent. Wrapping the dominant hand around the opponent's body, the fencer performing durchlauffen completes the action by throwing the opponent over their hip in a forward motion.

Fencers bind upon another and are too near to comfortably perform further work with weapons.

Blue releases grip with right hand and allows sword to hand behind back.

Blue takes small passing step with left foot and drives right arm into Red's torso.

Blue passes with right foot behind Red to disrupt structure and pulls Red forward, over hip.

Blue completes throw.

Abschneiden - Slicing Off

There are four slices a fencer can attack with. Two of these are over-slices which fall upon the opponent's wrists as they attempt to leave the bind. The other two under-slices press upward into an opponent's wrists as they raise their arms high while in the bind.

Over-Slices

When someone binds upon a fencer's sword and attempts to cut around the bind and strike to the opposite side, the fencer can perform a large step away from the attack. While stepping away from the cut, the fencer maintains the pressure they had in the bind but redirects this pressure laterally so the long-edge comes to rest over the opponent's wrists. This is not an arcing cut. The fencer performing the slice should simply follow their opponent's arms as they leave the bind and press the long-edge into the wrists. The two over-slices can be performed from both sides.

Fencers are bound upon one another.

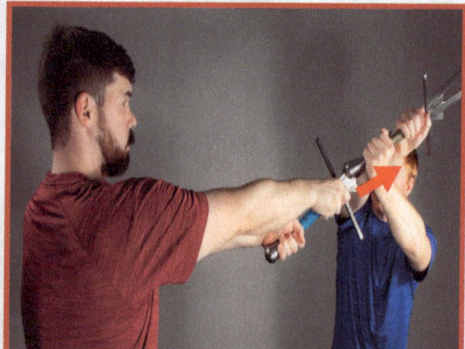

Red counters attempt to cut around by using slice.

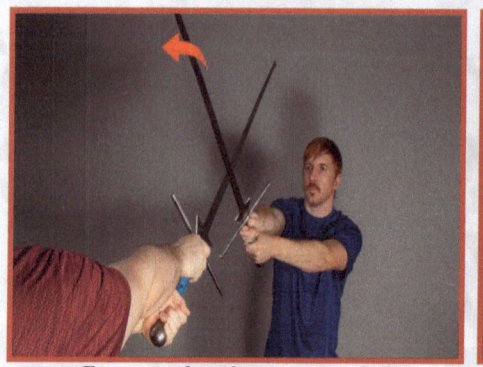

Fencers are bound upon one another.

Red counters attempt to cut around by using slice.

UNDER-SLICES

The under-slices are employed when someone binds upon a fencer's sword and attempts to overpower them by raising their hands high and rushing forward. The first under-slice can be employed from a bind when someone rushes forward into a fencer's non-dominant side. The fencer being overpowered responds by turning the sword so that the point is directed to their non-dominant side and pressing upward into the opponent's wrists using the long-edge. The other under-slice can be used when an opponent binds upon a fencer and attempts to overpower them on their dominant side. The counter to this is turning the sword so the point is directed to the fencer's dominant side and pressing upward into the opponent's wrists with the short-edge. Similarly to the over-slices, these attacks are delivered in tight motions.

Fencers are bound upon one another.

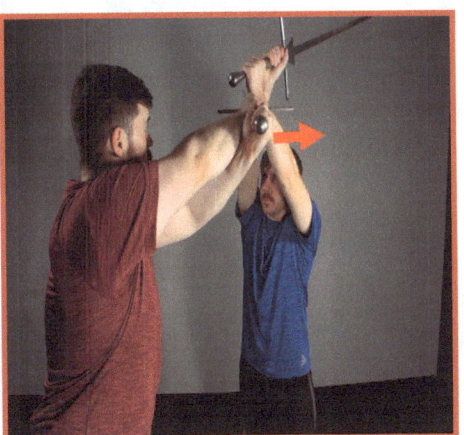

Red slices into wrists as Blue raises hands.

Fencers are bound upon one another.

Red slices into wrists as Blue raises hands.

HEND TRUCKEN - HAND PRESSING

Hend trucken involves turning an under-slice into an over-slice. Knowing how to quickly wind from an under-slice to an over-slice allows a fencer create opportunities in the krieg for followup attacks. When someone holds their hands high in the air and attempts to rush forward towards a fencer's non-dominant side, the fencer being charged can counter this by driving the long-edge into the opponent's arms. While stepping out towards their non-dominant side, the fencer maintains pressure upon the arms but begins to redirect it laterally so that they suppress the opponent's arms with the over-slice.

When someone similarly charges with their hands held high towards a fencer's dominant side, the fencer being charged counters by driving the false-edge into the opponent's arms. The fencer performing the slice steps to their dominant side while rolling the long-edge over the opponent's arms and suppresses them with the over-slice.

WHY SLICE?

It may seem more intuitive to perform a cut or thrust against a fencer preparing to overwhelm their opponent, but there is no guarantee either will stop the opponent from completing their attack. The under-slice in this scenario offers safety and control.

Blue rushes forward with hands raised high.

Red performs passing step with right foot and drives slice with false-edge.

Red performs triangle step while turning long-edge over Blue's wrists and forcing them down.

Hangen - Hangings

While attacks are typically delivered with extended arms from zufechten, it is often necessary to shorten the arms while attempting to find openings in the krieg where there is limited room to operate. There are four hangings and they are derived from pflugh and ochs. The two lower-hangings are pflugh from both sides while the upper-hangings are ochs from both sides.

Winden - Windings

Winding describes movement through the hangings while seeking openings to attack. Two possible windings are derived from each of the four hangers, making them eight in total. These eight windings are dictated by strong or soft pressure, and a fencer moves through the windings as they feel pressure in-the-moment. With each winding, a fencer attacks with a cut, thrust, or slice. Because there are three possible forms of attack which can be fenced from the eight windings, there are twenty-four attack variants.

Winding Thrusts

The RDL glosses offer an example of how winding can be conducted from the upper-openings. If a fencer is attacked to the upper opening on their non-dominant side, they can displace the attack with the short-edge while winding into the upper-hanger on their non-dominant side. With their arms raised high, the fencer can attempt to thrust below from the upper-hanging. If the opponent presses the point away, the fencer can wind into the opposite upper-hanging while yielding to the pressure and attempt another thrust.

If a fencer is attacked to the upper opening on their dominant side, they can displace the attack with the long-edge while winding into the upper-hanger on their dominant side. With their arms raised high, the fencer can attempt to thrust below from the upper-hanger. If the opponent presses the point away, the fencer can wind into the opposite upper-hanging while yielding to the pressure and attempt another thrust.

The above actions describe the four winding thrusts from the upper-openings. A fencer could attack in a similar manner while winding the point through the lower-hangings.

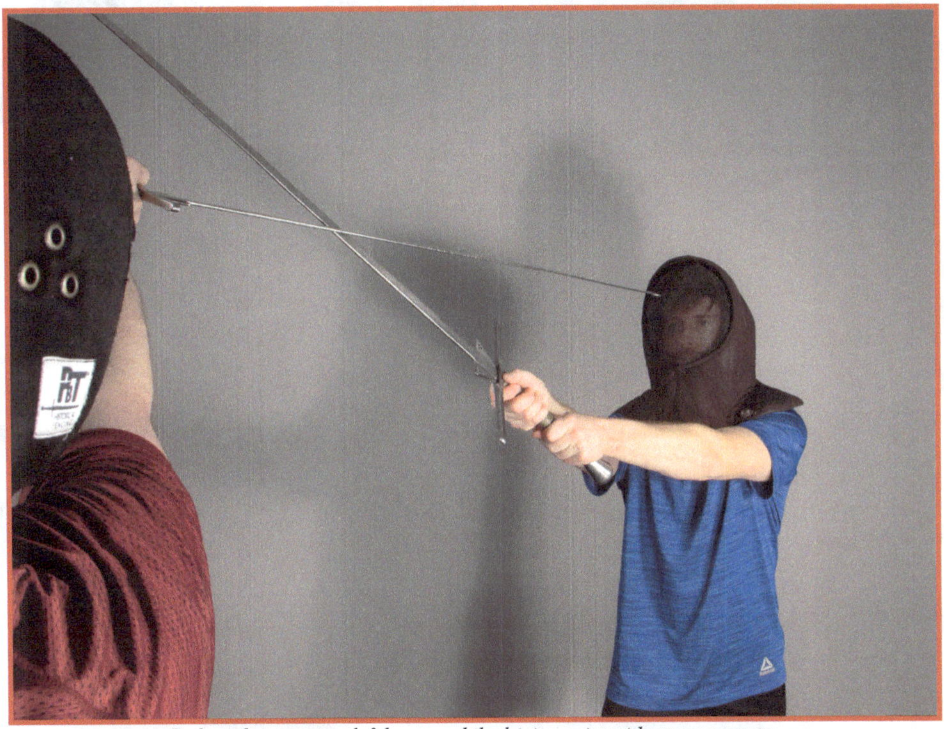

Red winds into upper-left hanger while driving point with strong pressure.

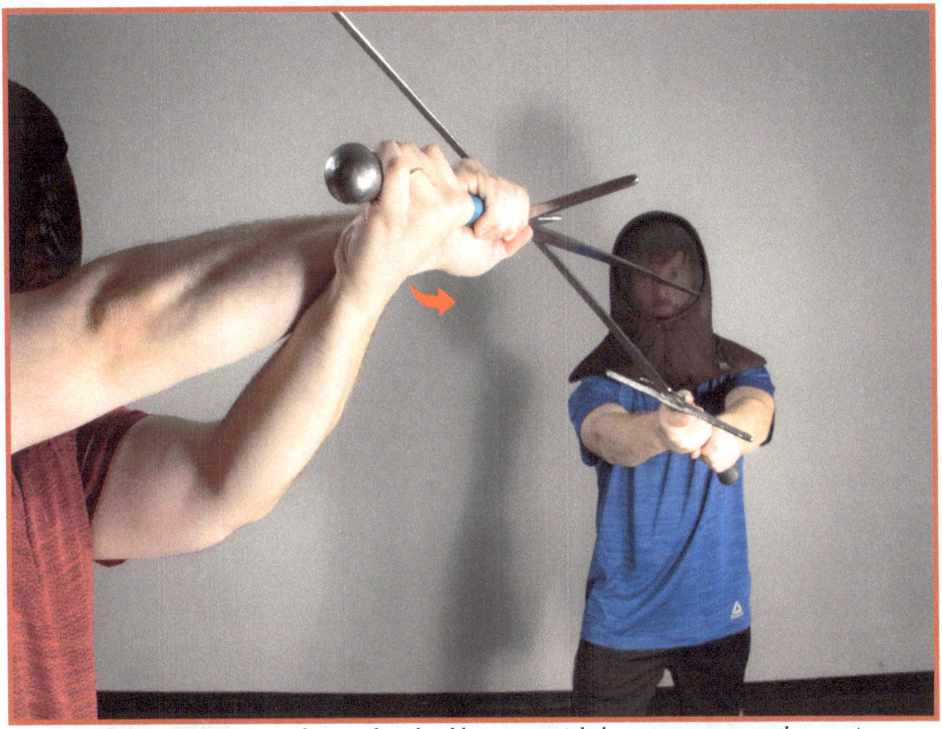

Red feels opponent resist with strength and yields to upper-right hanger to pursue another opening.

MUTIEREN - MUTATING

When a fencer performs an oberhau and their target parries it, they can attack with a *mutieren* if they feel that their opponent is soft in the bind. This attack is particularly effective when the opponent attempts to parry with the weak of their sword while moving into a high hanging, rather than attempting to parry with kron.

The fencer performing the mutieren winds their point to the outside of the opponent's sword while moving into a high hanging position. While pressing the opponent's weak down with the false-edge, the mutieren threatens the lower openings with the point.

WINDING CUTS

Meyer writes that the zwerchau's ability to threaten at close-distances is so significant that sword-fighting without it is like "half fencing." The zwerchau is capable of attacking at distances where long strikes cannot because the arms are shortened in the upper hangers. Winding cuts in close proximity to the target is therefore an effective way of stifling opponent's that only know how to cut by fully extending their arms.

DUPLIEREN - DOUBLING

When a fencer performs an oberhau and their opponent binds upon them strongly, they may attack with the *duplieren*. Immediately upon feeling that the opponent is firm in the bind, the fencer crosses their arms and cuts behind the opponent's sword, into their head.

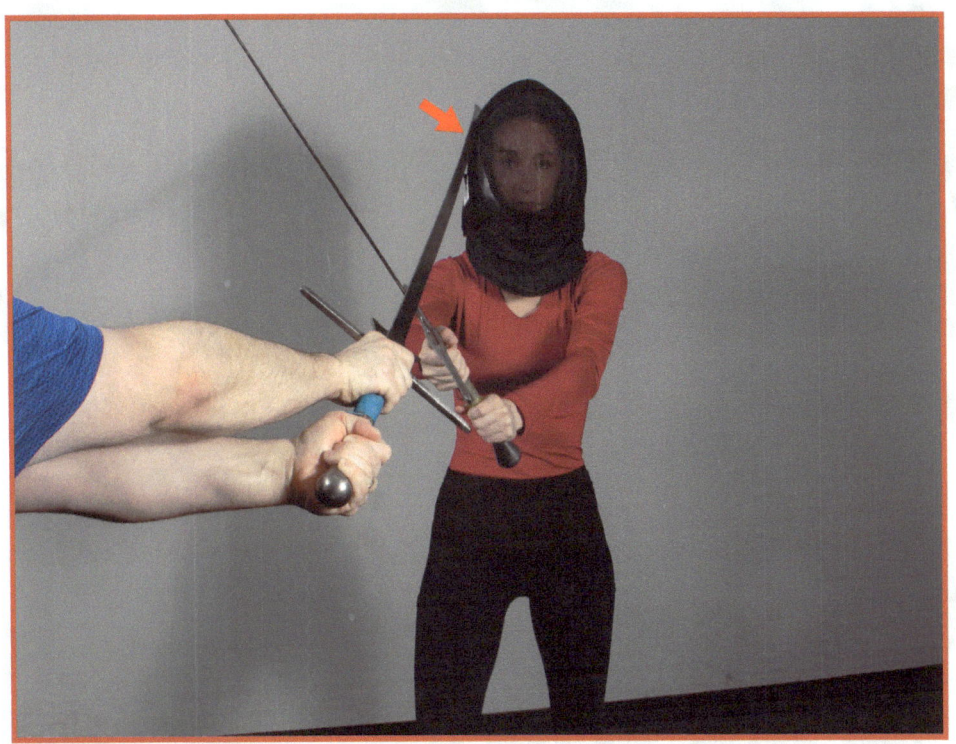

WINDING SLICES

The method of winding slices into the hangings has already been covered under abschneiden and hend trucken. The under-slices wind upwards into the upper hangings while the over-slices press down into the lower hangings.

> ### SIGNIFICANCE OF WINDING
> Winding attacks in the krieg from the shortened hangers is a critical component of Liechtenauer's art. The windings are the foundation of the art and all techniques and plays are derived from the windings. The MS 3227a suggests that the windings are unique to KdF and that "illegitimate masters" which only fence with extended arms dismiss the utility of the windings.

LANGENORT - LONGPOINT

Similar to postures within the vier leger, *langenort* is a guard which a fencer can adopt. Fencers can probe with the guard by standing with the non-dominant foot forward and extending the arms forward so the point threatens the opponent's upper-openings. While standing in this manner, a fencer attempts to determine the intention of their opponent so that they may counter them appropriately.

VIER LEGER VS LANGENORT

The vier leger are positions which deny the hands as an easy target, forcing the opponent to fence to deeper openings which a KdF fencer is well-trained to defend. While Danzig and Lew write that longpoint is a guard fencers can securely stand in, it is perhaps unwise to idle in the posture for too long with the hands exposed. This is supported by advice to set the point forward when the opponent is near and before the swords bind. It may be useful to think of longpoint as a provocation that is presented moments before an opponent prepares to close measure. With the point suddenly threatening near the upper openings, the opponent must make a quick decision which the fencer presenting longpoint is prepared to counter.

Longpoint Plays

When threatened with longpoint, an opponent is likely to react in a number of predictable ways. Assuming they do not react at all to the threatening point, the fencer presenting longpoint can simply pass forward and stab into the upper openings. If the opponent performs an oberhau, the fencer in longpoint can absetzen and thrust. If the opponent attempts to strike the fencer's sword, and not their person, the fencer in longpoint can durwechseln and thrust.

Fencers approaches opponent in vom tag.

Blue threatens with longpoint.

Red initiates oberhau. Blue performs a passing step while executing absetzen.

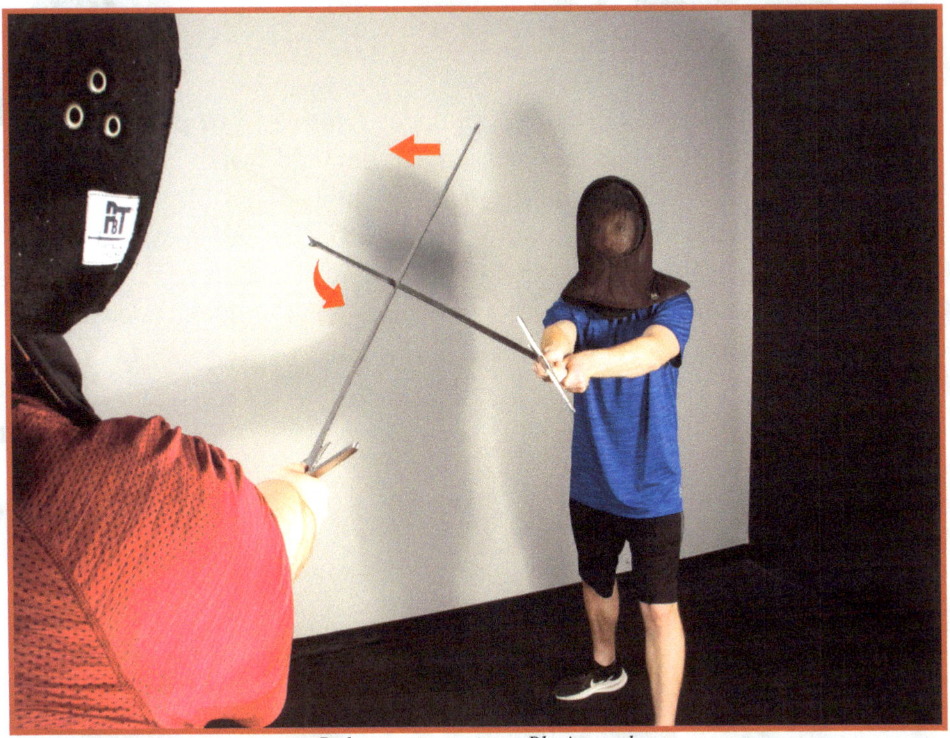

Red attempts to swipe at Blue's sword.

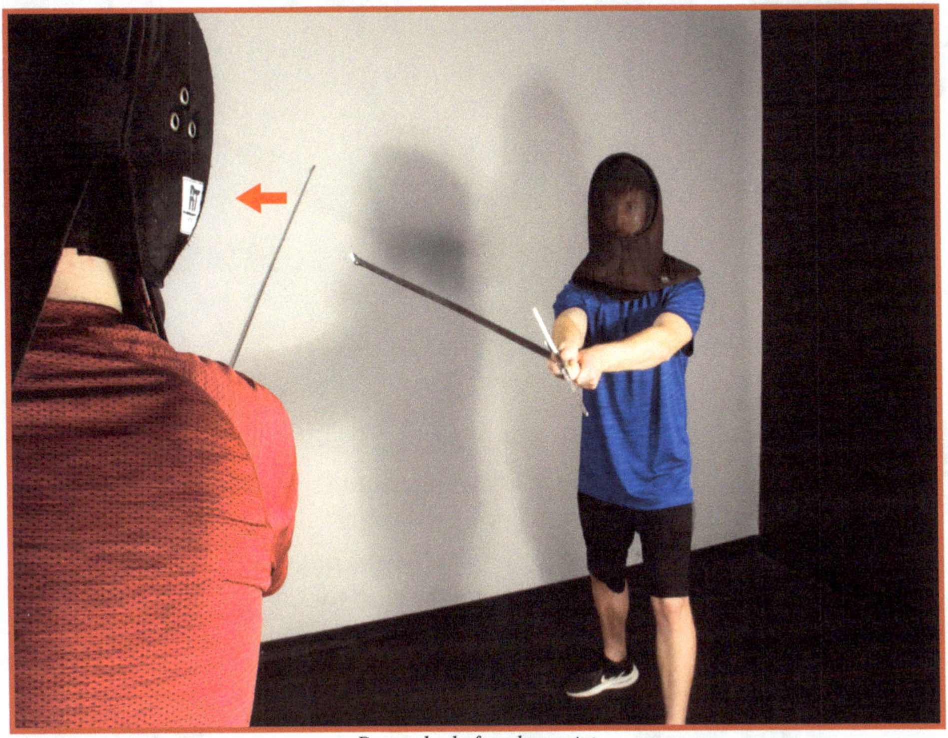

Durwechseln from longpoint.

SPRECHFENSTER - SPEAKING WINDOW

When a fencer has bound upon the opponent's sword while extending their arms in longpoint, they can feel pressures in the bind which inform them of their opponent's intentions in-the-moment they are performed. The fencer testing pressure in speaking window keeps their arms extended and their point threatening, while pressing firmly with the long-edge. Speaking window often results from a fencer performing a cut with extended arms which is parried by the opponent.

SPEAKING WINDOW PLAYS

If the opponent cuts around from the bind with an oberhau, the fencer in speaking window can drive their long-edge into the opponent's head while binding against their cut. The fencer in speaking window can similarly counter a zwerch by performing the slice as the opponent attempts to cut around the bind. If the opponent pulls their sword short in an effort to disengage from the bind and potentially durchwechseln, the fencer in speaking window can perform ansetzen. If the opponent remains still in the bind, a fencer in speaking window can perform a duplieren.

Fencers stand in speaking window.

Blue raises hand and prepares to cut around the bind with an oberhau.

Red drives long edge into Blue's head with a passing step before Blue can complete cut.

Fencers stand in speaking window.

Blue lifts hands and prepares to cut around bind with zwerchau.

Red drives long edge over Blue's wrists, slicing away the attack.

Fencers stand in speaking window.

Blue attempts to disengage under bind. Red performs ansetzen.

ADDITIONAL TECHNIQUES

The following techniques are either described within the early glosses or later German fencing writings. These techniques are valuable tools which can be applied in a variety of fencing situations.

SCHNAPPEN - SNAPPING

This is a useful technique for countering someone who wishes to strongly suppress a fencer's sword. *Schnappen* often occurs when a fencer performs an unterhau or extends the longpoint and their opponent then binds strongly to the sword rather than seeking an opening. A fencer performs schnappen by yielding to their opponent's pressure while rolling their pommel over their opponent's sword and striking them to the head with a cut.

Red performs unterhau. Blue falls onto the cut with long edge.

Blue suppresses cut with strong lateral pressure. Blue is not threatening with point.

Red snaps around the bind with a passing step, executing an oberhau to upper opening.

REISSEN - WRENCHING

Wrenching is often used to describe actions where the hilt of the sword is used in a violent pulling or shoving motion. This is a useful technique against opponents who parry blows without attempting to threaten. The wrench controls the opponent and prevents further parrying.

WRENCHING WITH POMMEL

If an opponent parries a strike, the attacker can roll their pommel over the defender's wrists and pull downward. The fencer wrenching downward uses the same motion to strike their opponent in the head with the long-edge.

> This method of wrenching works best when the fencer is not overly focused on pulling the opponent's arms down. Instead, the fencer executing the wrench should think about performing an oberhau behind the opponent's sword. While cutting downward with the pommel over the opponent's wrists, the attacker will wrench their target in the same motion, thus preventing them from defending.

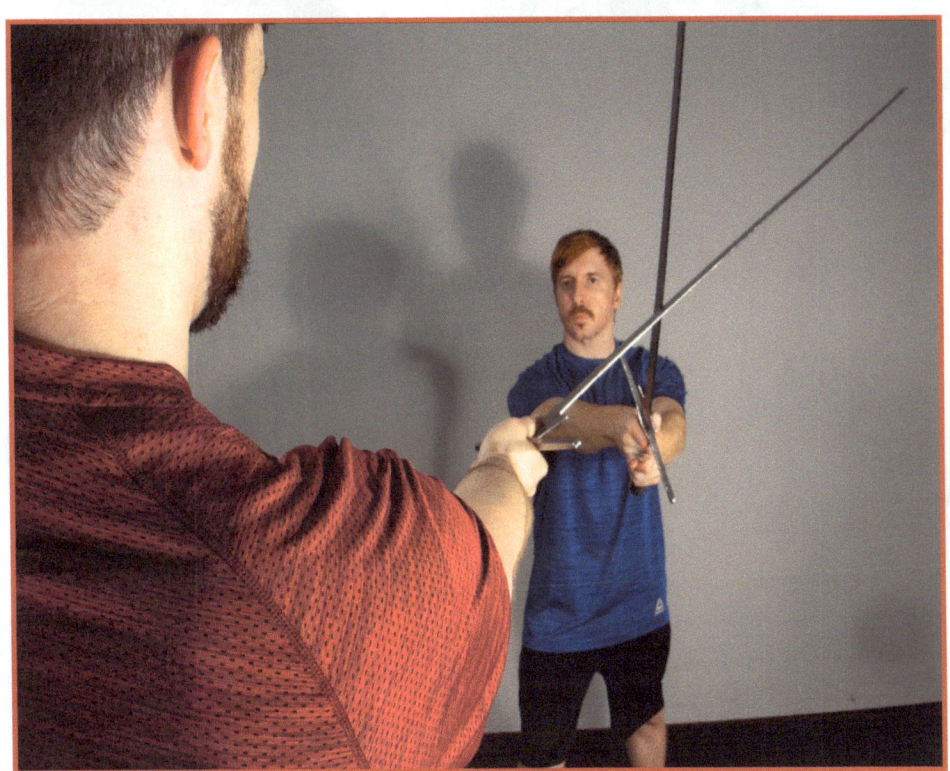

Red performs oberhau. Blue displaces with lateral parry.

Red rolls pommel over Blue's wrists.

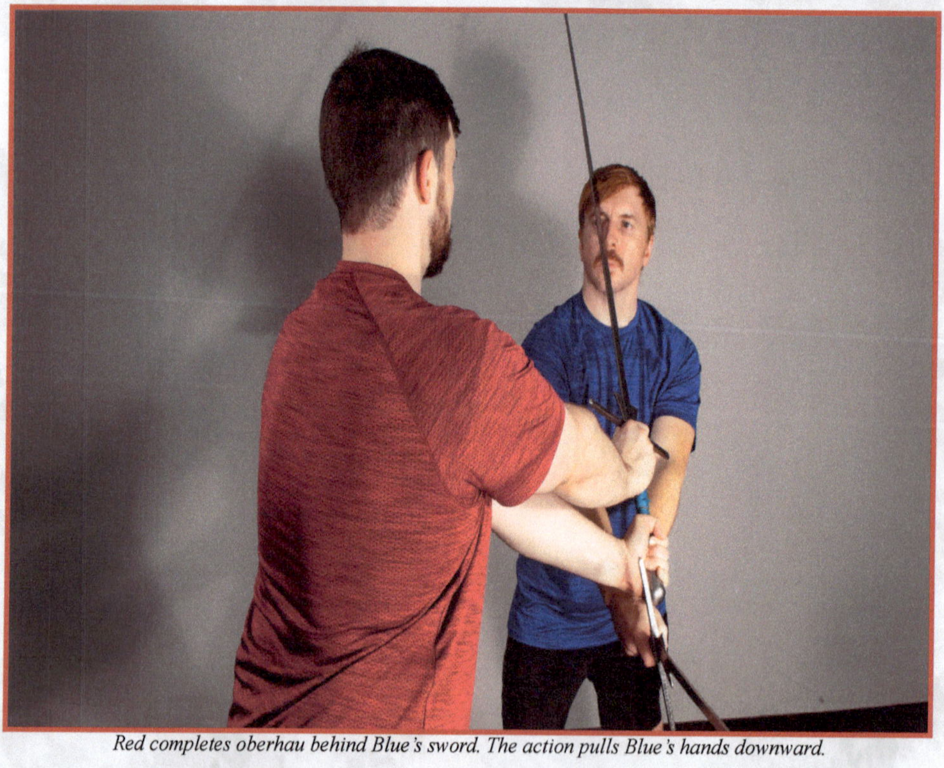

Red completes oberhau behind Blue's sword. The action pulls Blue's hands downward.

WRENCHING WITH CROSSGUARD

If an opponent parries a strike, the a attacker can shove the opponent's sword aside with the crossguard. In the same motion as the wrench, the attacker executes another strike to the opposite side.

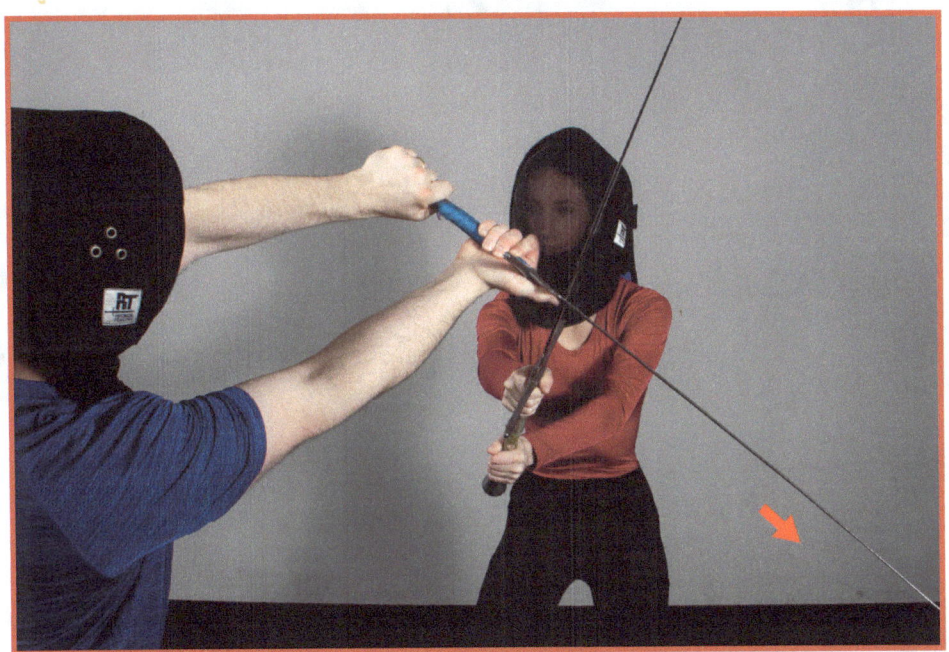

Blue's zwerchau has been parried by Red. Blue shoves Red's sword away and downward.

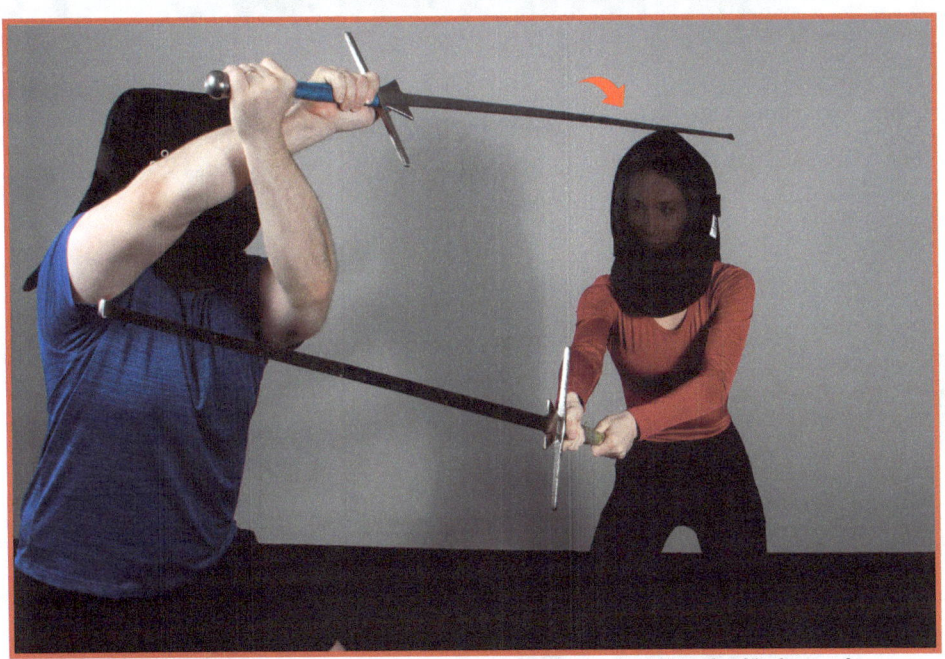

Blue follows motion of the wrench and performs zwerchau to the other side of Red's sword.

RINGEN IM SCHWERT - WRESTLING TO THE SWORD

Alternatively to performing throws while grappling, a fencer can seek to control their opponent's weapon. Wrestling to the sword, rather than the body, is ideal when the opponent's hands are too low for durchlauffen to be viable. An example of how to control the opponent's sword is by grabbing their hilt. When a fencer is near their opponent in the bind, they can invert their non-dominant hand and grab the opponent's hilt, between their hands. With the opponent's hilt in hand, the fencer violently jerks the sword to their non-dominant side and cuts them upon the head with a one-handed blow.

MAINTAINING PRESSURE

When preparing to grapple, it is important to maintain forward pressure in the bind. This is true both before and after the non-dominant hand has been removed from grip. This prevents the opponent from escaping the bind or wrenching the grappler aside.

Fencer's enter close bind where sword work is difficult to execute. Hands are too low for durchlauffen.

Red inverts left hand and grab's Blue's hilt. Red maintains pressure in bind with right hand.

Red pulls Blue towards their left side, redirecting Blue's strong forward pressure.

SWORD-TAKING

Opportunities to disarm an opponent may also arise from wrestling. An example of this is when a fencer inverts their non-dominant hand and grabs both swords in a bind, near the middle of each blade. The fencer then moves their dominant hand under the bind while raising the pommel forward, over their opponent's arms. With a violent pull upwards and towards their dominant side, the fencer wrenches the sword free from their opponent's grasp and completes the disarm.

Fencers are bound upon one another. Blue inverts left hand and grasps both blades in the middle.

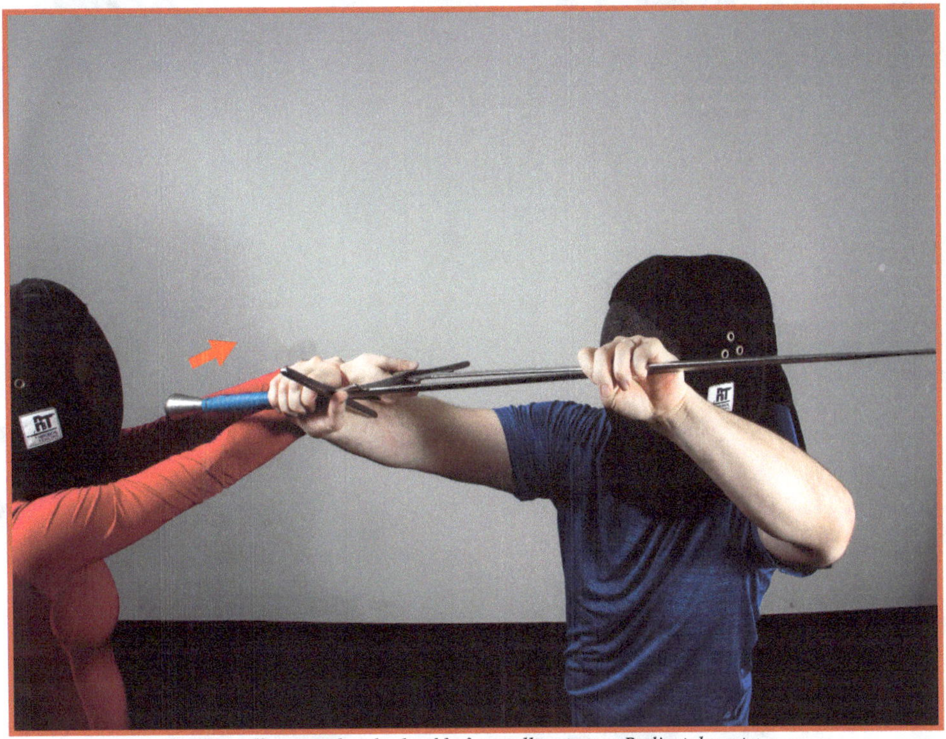

Blue rolls pommel under bind before pulling it over Red's right wrist.

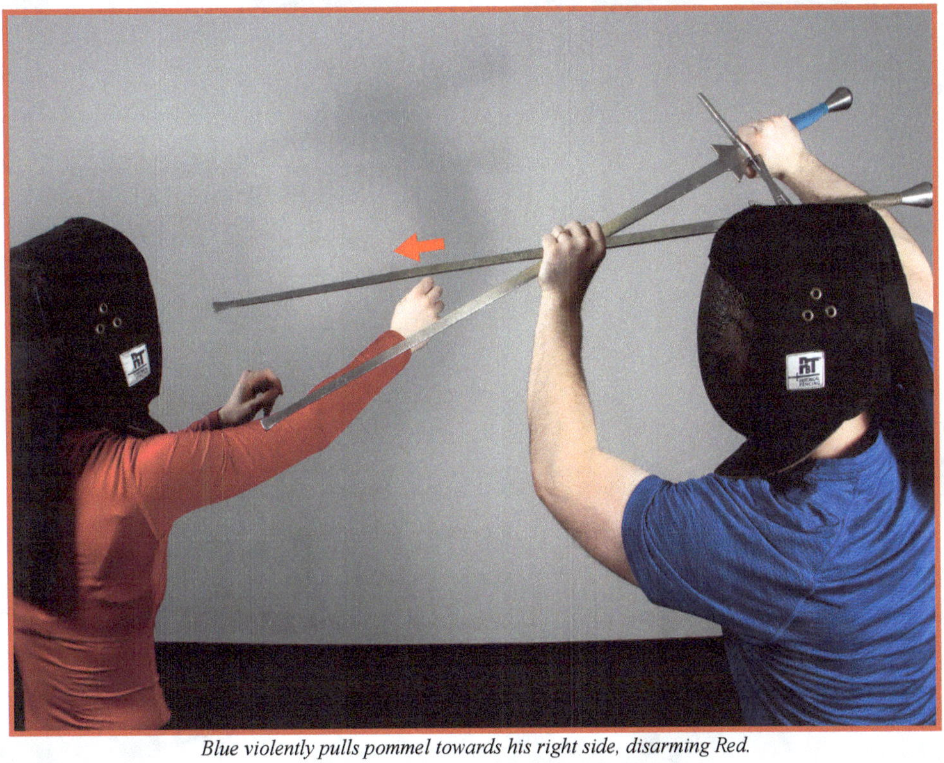

Blue violently pulls pommel towards his right side, disarming Red.

ELBOW PUSH

Disarms and hilt manipulations can be difficult to perform. A simpler grappling technique involves pressing the opponent laterally by the elbow. It is important for the fencer to press the elbow while driving with the entire body, rather than relying solely on the arm to generate power. The fencer should also strive to control their opponent by the elbow while stepping to the side, rather than trying to out muscle the target while squaring up with them.

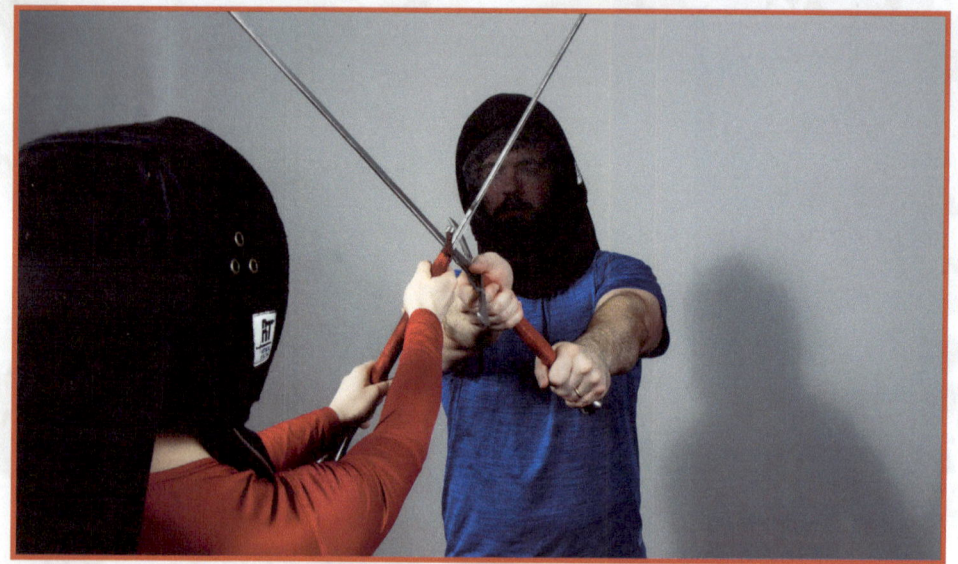

Fencers are bound upon one another.

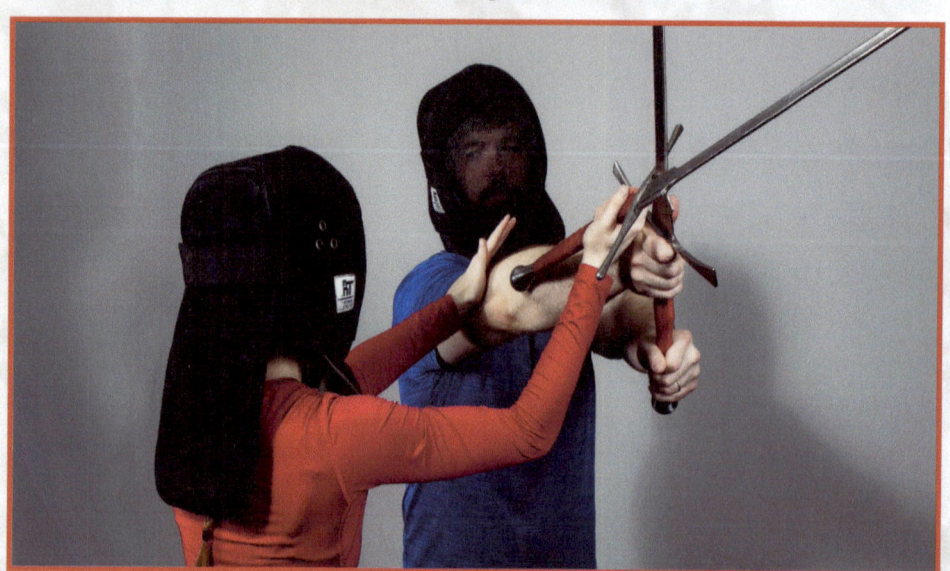

Red steps towards Blue's right side while placing left hand on elbow.

Red extends left arm and continues pressing with entire body. Red is chambered to cut or pommel Blue.

HALF-SWORDING

While fencing from the shortened, or half-sword, is more often done against opponents wearing armor, it can be effectively used against unarmored opponents as well. Once bound upon an opponent's sword, a fencer can cut around to the other side of the bind and while doing so, grip the middle of their blade with the non-dominant had. With the added leverage provided from half-swording, the fencer can easily seek the upper openings with the point.

Fencers are bound upon one another.

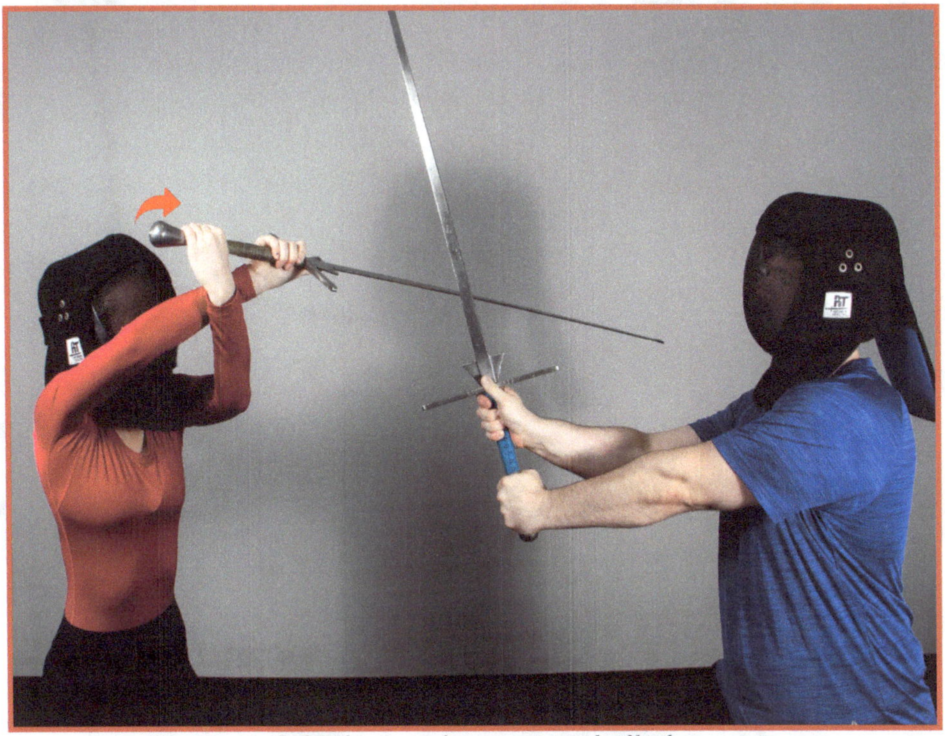

Red performs zwerchau to opposite side of bind.

Blue parries the zwerchau. Red grasps end of their own sword with left hand and thrusts.

GAYSZLEN - WHIP/SCOURGE

Hans Talhoffer depicts a fencer performing a one-handed cut to the opponent's legs while holding the pommel with the non-dominant hand. No commentary or context is provided about this unique cut. The additional reach offered by gripping the sword near the pommel allows a fencer to somewhat cheat uberlauffen.

WHEN TO WHIP

While the gayszlen's reach does provide some safety while attacking the lower openings from zufechten, fencers should avoid recklessly defying the principles of uberlauffen. An aggressive opponent who is actively attempting to fence to the upper openings is not an ideal target of the gayzlen as the likelihood of being struck while executing the one-handed attack is great. The gayszeln works best when the target is flat-footed, backpedaling, or is looking to parry instead of seeking the vor.

Mittelhau - Middle Strike

The *mittelhau* is a horizontal strike performed with extended arms. The cut is delivered with the long-edge to both of the upper openings. Meyer advises this cut functions similarly to the zornhau.

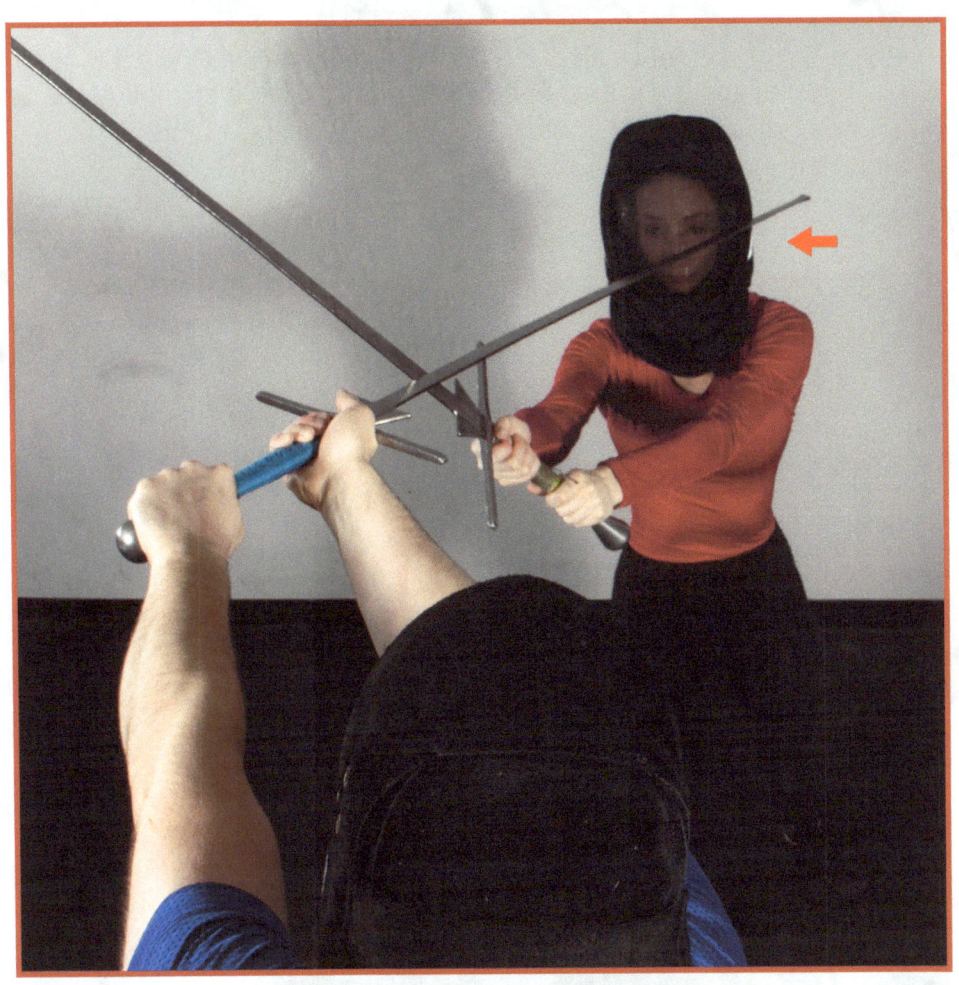

Principal Strike

While Meyer considers the mittelhau to be one of his four principal strikes, the cut does not appear in the early RDL sources. Furthermore, Meyer contradicts earlier sources as he claims there are four principal strikes from which all cuts derive (oberhau, zornhau, mittlehau, and unterhau), while the MS 3227a advises all cuts are derived from the oberhau and unterhau. Because this book focuses on the early KdF sources, the mittelhau was not included with the basic cuts.

MITTELHAU RISKS

Delivering a mittelhau to the opponent's torso carries significant risk. If executed from the onset, the fencer's upper openings are vulnerable due to uberlauffen. In the krieg, the low mittelhau leaves a fencer exposed, and is particularly dangerous if poorly timed with a simultaneous attack from the opponent. Striking to the body is safest when the opponent is raising their hands high while attempting to parry. If striking the torso in this manner, a fencer should immediately withdraw from the target following the cut.

Blue executes low mittelhau while in the krieg. Red simultaneously strikes Blue in the head.

HANGENORT - HANGING POINT

The hanging point is a posture similar to ochs, but the fencer allows the sword to hang so the point is directed towards the ground. A fencer can use the hanging point to parry descending blows, deflecting them off the ramp they create with their blade. Following the deflection, a fencer can quickly followup with a long-edge strike.

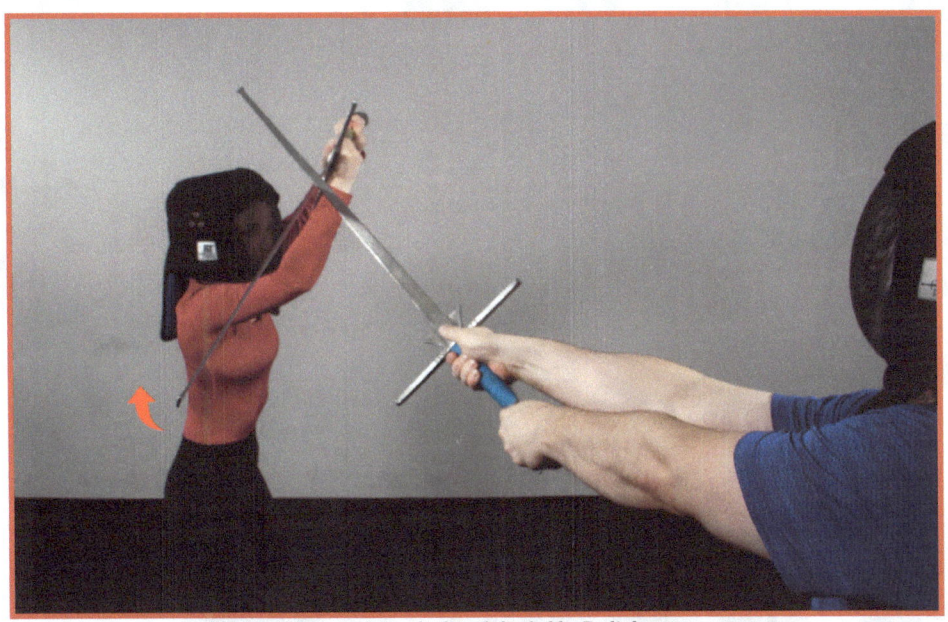

Blue performs oberhau which is defended by Red's hanging point.

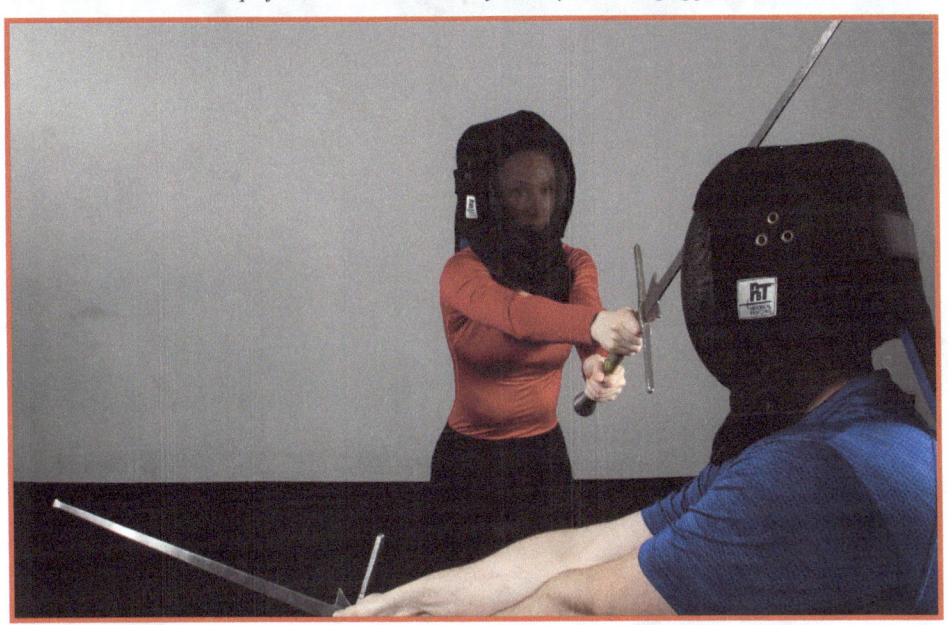

Red deflects Blue's blows before stepping with a cut.

STREYCHEN - SWEEPS

While not mentioned in the Zettel[17], the sweeps are described in various German fencing manuals alongside the Zettel and its teachings. These short-edge strikes begin from low positions where the point is near the ground and end with the point directed towards the sky. Sweeps can be performed from both sides with either foot forward, but are most effective when performed from a fencers non-dominant side.

EISENPORT & NEBENHUT - IRON GATE AND NEAR GUARD

Eisenport and nebenhut are two alber-like postures from which a fencer can execute sweeps. Nebenhut is held with the point directed behind the fencer and near the ground. Eisenport is held with the point directed to the ground, somewhat forward and to either side of the fencer.

Nebehnhut

Eisenport

Sweeping Deflections

Deflecting attacks is the most common function of the sweeps. Sweeps provide opportunities for the defender to quickly find openings following the deflection. If a fencer feels that their opponent is soft in the bind while sweeping upwards, they can immediately drop down with an oberhau to the upper openings. If the opponent attempts to suppress a sweep with strength, the fencer performing the sweep can use schnappen to strike the upper openings. An opponent which binds firmly against a sweep — but is not overly aggressive in pressing against it — is vulnerable to duplieren.

Powering the Sweeps

Sweeps mechanically function similarly to other cuts. Fencers performing a sweep should keep their arms lengthened with proper structure throughout. Turning the hips while activating the core drives the sweep upward.

Red approaches in vom tag. Blue stands in eisenport.

Blue sweeps away Red's oberhau. The sweep significantly displaces Red's sword.

Blue cuts down into the left side of Red's head with long-edge.

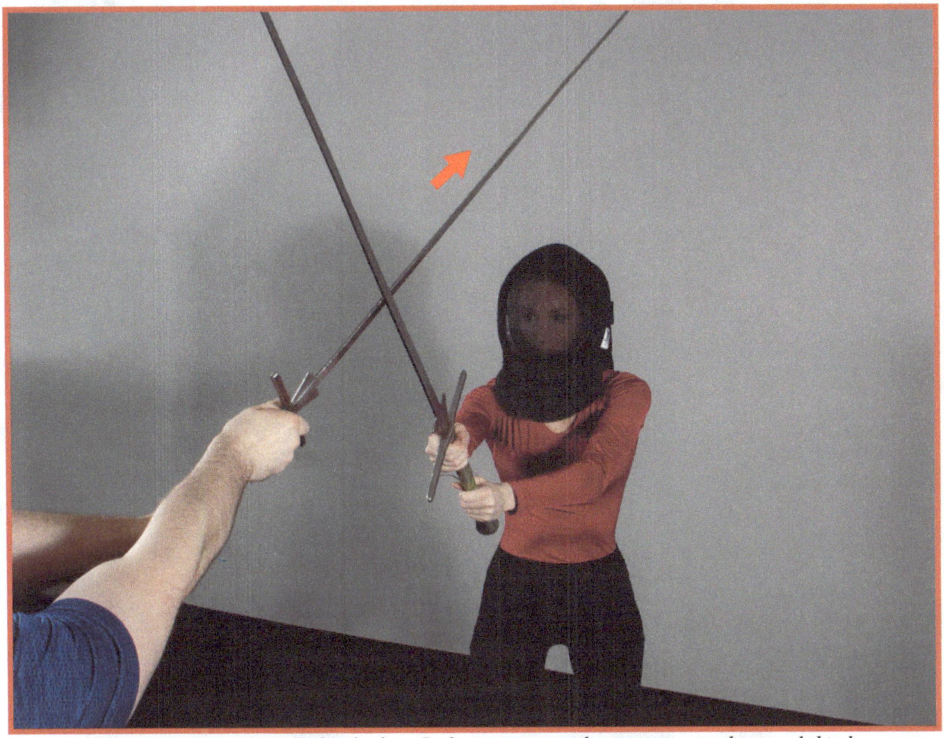

Blue sweeps up against Red's oberhau. Red maintains steady pressure once the swords bind.

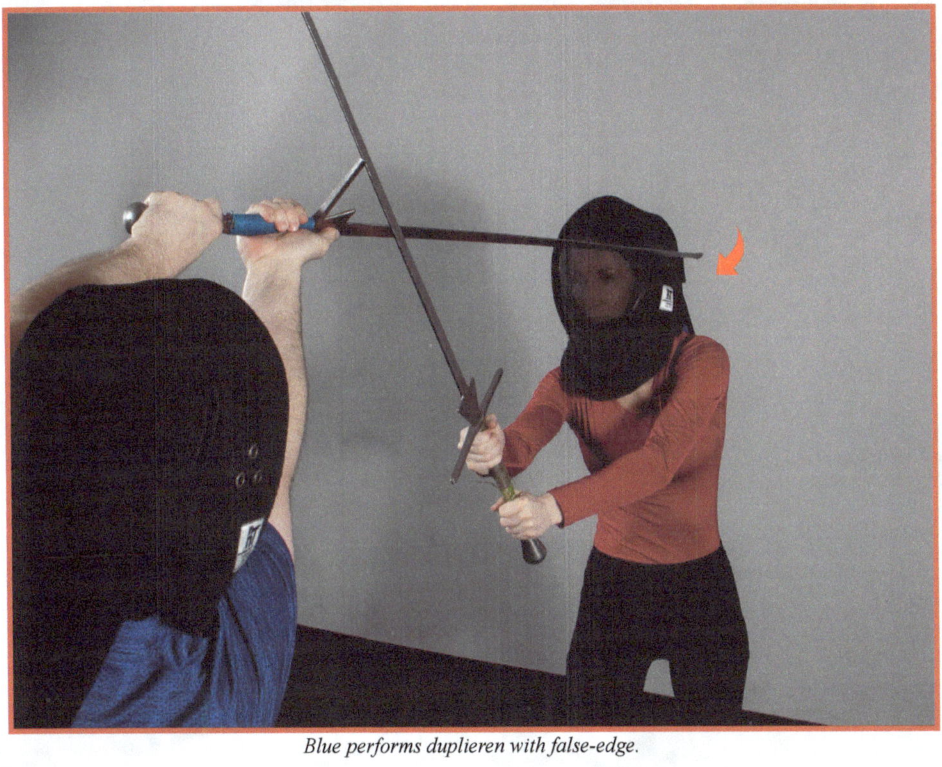

Blue performs duplieren with false-edge.

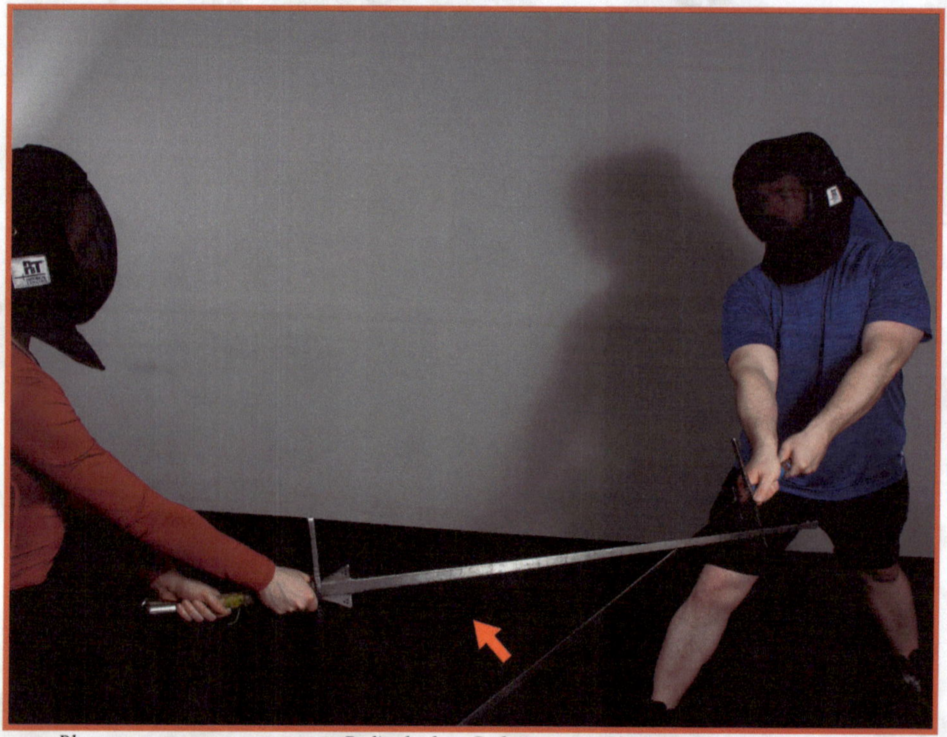

Blue attempts to sweep up against Red's oberhau. Red suppresses Blue's sword with strong pressure.

Blue yields to pressure and snaps around bind with false-edge and triangle step.

STURTZHAU - PLUNGING STRIKE

Setting the non-dominant foot forward, a fencer motions as though they intend to perform an oberhau from their dominant shoulder. If the defender raises their hands high to intercept the strike, the attacker can wind into their upper-right hanging and thrust beneath the defender's hands.[18]

WRENCHING FROM STURTZHAU

If a fencer motions as though they will perform an oberhau and realizes their opponent will counter with an descending cut of their own, they can invert their hands and strike with a descending false-edge cut. Pulling the short-edge downwards and toward their non-dominant side, the fencer performing the sturzhau can wrench their opponent's sword away and quickly follow up with another cut.[19]

Blue wrenches downwards with false-edge upon Red's sword.

Blue completes wrench with a passing step and cuts with long-edge before Red can recover.

PRELLHAU - BOUNCE STRIKE

The *prellhau* is a strike performed with the flat. This attack serves a variety of potential functions. When delivered to the head, a prellhau could presumably debilitate a target with less lethality than a strike performed with the edge.

PRELLHAU AS A PROVOCATION

Striking an opponent with the flat may provoke them to swipe away the attack, exposing their opposite openings to further attack.[20]

Blue strikes with flat towards Red's head.

Red attempts to defend strike while Blue cuts around to other side of Red's sword.

Prellhau as a Deflection

Hitting an opponent's sword with a flat strike can deflect an incoming attack. The fencer performing the prellhau can immediately attack after bouncing off the opponent's blade.

Blue strikes Red's sword with Prellhauw.

Blue rolls sword over head and performs another flat strike to the side of Red's head.

ZIRCKEL - CIRCLE

Joachim Meyer and Andre Paurenfeyndt describe a false-edge cut delivered from a bind, which is directed to the opponent's upper-right opening. Meyer refers to this handwork as the *circle* and explains it is derived from the krumphau,[21] while Paurenfeyndt describes the attack as a *crown cut.* Cutting in this manner is another example of how to wind cuts in the krieg with shortened arms when there is little room to operate.

  The circle can be performed after a fencer has bound onto their opponent with a long-edge cut. Rather than cutting around to the other side with the long-edge which is already pressing in the bind, the fencer can instead cross their hands and descend the false-edge into the opposite upper-opening. Meyer advises this cut can strike very hard, or be used as a provocation to force the opponent to chase laterally with the parry. Whether the circle hits or not, the fencer executing it can quickly drop their long edge back into the original opening.

Fencers are bound upon one another.

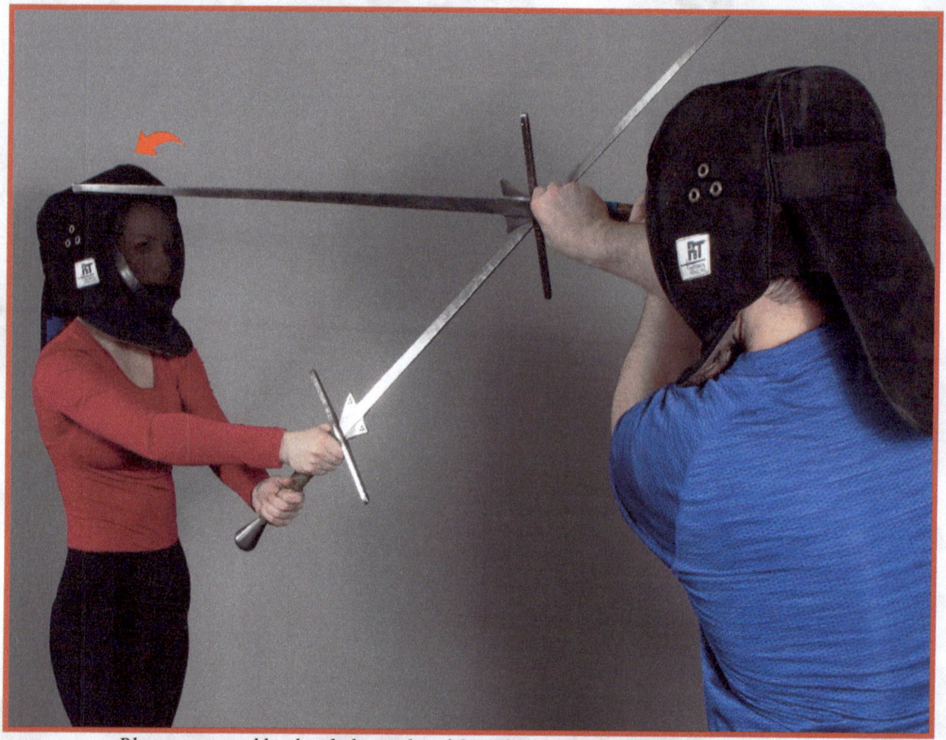

Blue cuts around bind with descending false-edge cut directed toward Red's right ear.

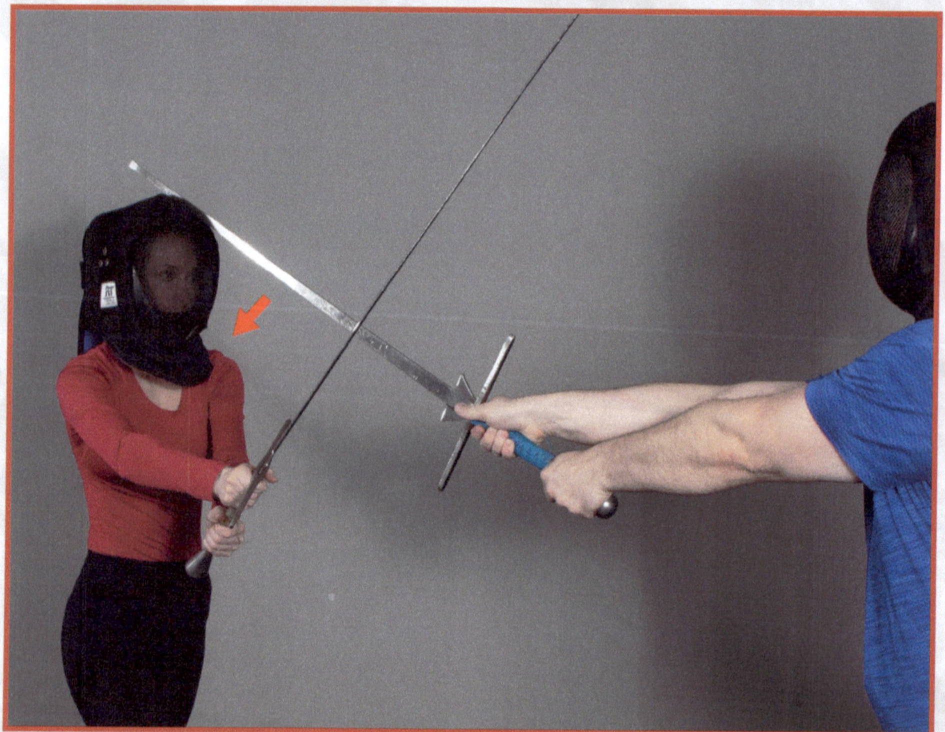

As red presses laterally to defend cut, Blue rolls false-edge cut into an oberhau on other side of Red's sword.

SUMMATION

Rather than waiting for the opponent to attack and relying on defense, fencers should be actively trying to apply pressure with their offense. The goal of the fight is to force an opponent to constantly defend in such a manner that they are limited in their ability to commit offensive actions of their own. Understanding the relationship between the vier leger and the vier versetzen allows a fencer to choose attacks appropriately. Reckless aggression is unwise however, as it may offer nachreissen opportunities to the opponent.

Seeking the upper openings with the point is a reliable way to attack from the onset, as attempting to deliver a committed cut can possibly result in an attack which is too short and ceases to threaten. When not racing-after opportunities offered by the opponent, attacks performed from the onset are most often successful when preceded with some manner of misdirection, such as a feler, durwechseln, or zucken.

If forced to parry or otherwise displace an opponent's attack, a fencer should seek to immediately turn that defense into offense. Absetzen allows a fencer to quickly seize the vor away from their opponent with point. The master strikes are also a useful way to harm the opponent while protecting oneself from attack.

While fencing with extended arms is ideal for reaching an opponent in the onset, it may be difficult for a fencer to attack in the same manner while pressed in the krieg. Winding attacks from the four hangings allows a fencer to shorten their arms and seek openings even when near their opponent. Opponents who are only comfortable fencing with long strikes and extended arms can find winding attacks unfamiliar and difficult to defend, providing a possible advantage to a KdF fencer.

When faced with an opponent who seeks to overpower others with strength and aggression, fencers may need to rely on slicing and wrestling. Under-slices can lead to over-slices and over-slices can offer many opportunities to harm the opponent. Durchlauffen may be necessary when two fencers are too near for sword-work.

From the bind, a fencer should allow pressure to dictate their decisions. Fencers should not be impetuous or reckless with their attacks, instead attacking in-the-moment they feel where the next opening presents itself. Similarly, a fencer should not idle in a bind for too long and potentially lose the vor.

GLOSSARY / INDEX

Abnehmen - Taking Off (page 33)
Lifting off the opponent's sword to cut around the bind. Likely related to the concept of zucken.

Abschneiden - Slicing Off (page 67)
Pressing with the edge to control the opponent.

Absetzen - Setting Off (page 56)
Parrying while maintaining a forward point.

Advancing Step (page 11)
Stepping forward with the lead foot. Rear foot may or may not follow.

Alber - Fool (page 51)
One of the four basic guards (vier leger). Arms are extended downward with point near the ground.

Ansetzen - Setting On (page 55)
Thrusting into upper opening as opponent lowers sword.

Bind (page 7)
Describes two swords making contact with one another.

Drei Wunder - Three Wounders (page 15)
Cutting, stabbing, slicing.

Duplieren - Doubling (page 74)
A cut delivered behind the opponent's sword.

Durchlauffen - Running Through (page 64)
A form of wrestling which typically involves throwing the opponent.

Durwechseln - Change Through (page 58)
Disengaging beneath the opponent's sword.

Einschiessen - Shooting In (page 21)
Often used to describe a thrust which is initiated from a bind when the opponent is offering soft pressure.

Eisenport - Iron Gate (page 98)
A retracted variation of alber. Used to set up streychen (sweeps).

Federschwert (Feder) - Feather Sword (page 5)
A training sword which features a narrow blade profile.

Feler - Failer (page 41)
An attack which is pulled short of its target, functioning as a feint.

Gayszlen - Whip/Scourge (page 97)
A one-handed cut to the opponent's legs.

Half-sword (page 95)
A sword which is gripped with one hand on the blade and the other on the hilt.

Hangenort - Hanging Point (page 100)
A posture similar to ochs, though the point is directed downward.

Hau - Strike/cut (page 15)
An attack that relies on the sword's rotation to deliver a percussive or severing blow.

Haupstucke - Chief Pieces (page 4)
These are the principal techniques and concept which govern KdF fencing. They include: zornhau, krumphau, zwerchau, schielhau, scheitelhau, vier leger, vier versetzen, nachreissen, uberlauffen, absetzen, durwechseln, zucken, durchlauffen, abschneiden, hend trucken, hangen, winden.

Hend Trucken - Hand Pressing (page 69)
Slices delivered to the opponent's wrists.

Hengen - Hanging (page 71)
The four hangers are ochs (both sides) and pflugh (both sides).

Indes - In-the-Moment (page 6)
Describes the small window of time (instantaneous) where decisions are made based on pressure in the bind.

Kunst des Fechtens (KdF) - Art of Fighting (page 2)
A martial art attributed to Johannes Liechtenauer.

Krieg - War (page 6)
This distance where a fencer can reach their target.

Krumphau - Crooked Cut (page 34)
One of the five master cuts. When performed from the dominant side, the cut is performed with crossed hands. Counters ochs.

Langenort - Longpoint (page 75)
A guard which is adopted by fully extending the arms, and the point, towards the target.

Long-Edge / True-Edge (page 5)
When the longsword is held in a standard grip, this is the edge which faces away from the fencer.

Master Strikes / Five Strikes (page 30)
Zornhau, krumphau, zwerchau, schielhau, scheitelhau

Mittelhau - Middle Cut (page 98)
A long-edge cut delivered with extended ams and which travels along a horizontal line.

Mutieren - Mutating (page 73)
Winding into a high hanger against opponent's soft pressure to set up a thrust to lower oppening.

Nach - After (page 6)
Fencer is reacting to opponent, rather than taking the initiative. A fencer operating in the nach is one threatened by an attack which they must defend against.

Nachreissen - Racing After (page 54)
Attacking an opening immediately after the opponent presents it.

Nachslag - After Strike (page 6)
Followup attack which forces opponent to continue focusing on defense after defending preceding attack.

Oberhau - Over Cut (page 15)
A descending cut.

Ochs - Ox (page 50)
One of the four basic guards (vier leger). Hands are held high while the point is directed towards opponent.

Passing Step (page 11)
The rear foot passes the lead foot.

Pflugh - Plow (page 51)
One of the four basic guards (vier leger). Hands are held low while the point is directed towards opponent.

Prellhau - Bounce Strike (page 108)
A blow delivered with the blade's flat.

Reissen - Wrenching (page 86)
Using the sword's hilt to manipulate an opponent or their weapon.

Ringen - Wrestling (page 64 & 89)
Can include wrestling to the body (durchlauffen) and wrestling to the sword (ringen im schwert).

Scheitelhau / Scheitler - Parting Cut (page 48)
One of the five master cuts. Long-edge descending cut which counters alber.

Schielhau / Schiller - Squint Cut (page 46)
One of the five master cuts. False-edge descending cut which counters pflugh.

Schnitt - Slice (page 22)
An attack which involves pressing a target with the sword's edge.

Schnappen - Snapping (page 84)
Yielding to hard lateral pressure while cutting around a bind.

Short-Edge / False-Edge (page 5)
This is the edge which faces towards the fencer.

Schrankhut - Barrier Guard (page 35)
This position is where a krumphau may be initiated and where all krumphau terminate their travel.

Sprechfenster - Speaking Window (page 79)
A variation of langenort that describes pressing into the opponent's blade while directing the point towards them.

Stich - Thust/Stab (page 20)
An attack made with the sword's point.

Sturtzhau - Plunge Strike (page 106)
A descending false-edge strike performed with crossed arms. Often used to set up a thrust.

Streychen - Sweeps (page 101)
Ascending false-edge strikes that primarily serve to deflect attacks.

Triangle step (page 12)
The rear foot travels laterally, behind the lead foot.

Uberlauffen - Over Running (page 55)
A concept which explains how the upper openings may be reached before the lower openings.

Unterhau - Unter Cut (page 18)
An ascending cut.

Vier Leger - Four Positions (page 50)
Alber, pflugh, ochs, vom tag.

Vier Versetzen - Four Displacements (page 53)
Krumphau, zwerchau, schielhau, scheitelhau.

Vom Tag - From the Day (page 52)
One of the four basic guards (vier leger). Hands are held high, near the shoulder or over the head, and chambered to cut.

Vor - Before (page 6)
Fencer seizes the initiative and threatens opponent so they must defend.

Vorslag - Before Strike (page 6)
The first strike performed from zufechten which draws a fencer into the krieg where they would ideally be capable of performing a series of nachslag.

Winden - Winding (page 71)
Describes moving into one of the four hangers, often in conjunction with an attack.

Zettel - Recital (pages 2, 23-29)
The cryptic poem attributed to Johannes Liechtenauer which preserves the core principals of his martial art.

Zirckel - Circle (page 110)
Handwork which is derived from the krumphau. Often manifests in a descending false-edge strike with crossed arms.

Zornhau - Wrath Cut (pages 30)
One of the five master cuts. Long-edge descending cut which counters oberhau.

Zornort - Wrath Point (page 31)
A thrust which follows the zornhau.

Zwerchau - Thwart Cut (page 38)
One of the five master cuts. When performed from the dominant side, this is a lateral false-edge strike. Counters vom tag.

Zucken - Pulling (page 59)
Disengaging above the opponent's sword.

Zufechten - Onset (page 6)
The distance where both fencers are maneuvering and preparing to attack. Fencers are incapable of striking from this distance without performing a passing step.

ENDNOTES

1   Jeffrey Forgeng, *The Art of Combat* (Yorkshire: Greenhill Books, 2006), 68

2   Michael Chidester, *The Long Sword Gloss of GNM Manuscript 3227a* (Somerville: HEMA Bookshelf, LLC, 2021), 43

3   Jeffrey Forgeng, *The Art of Combat* (Yorkshire: Greenhill Books, 2006), 68

4   Stephen Cheney, *Ringeck Danzig Lew (RDL) Longsword* (Coppell: Bucks Historical Fencing / Philadelphia European Martial Arts Collective, 2020), 5-6

5   Jeffrey Forgeng, *The Art of Combat* (Yorkshire: Greenhill Books, 2006), 69

6   Michael Chidester, *The Long Sword Gloss of GNM Manuscript 3227a* (Somerville: HEMA Bookshelf, LLC, 2021), 33-34

7   The modern colloquial, "thumb grip," does not appear in any historical sources. "Reversed grip" likewise does not appear in any sources, though it was inspired by the "reversed cuts" which Meyer defines as strikes performed with short edge, flat, or some angle. While these reversed cuts would not necessarily be performed with the thumb extended along the flat, they would often be performed with the sword reoriented in the hand in some manner. See Jeffrey Forgeng, *The Art of Combat* (Yorkshire: Greenhill Books, 2006), 56.

8   Michael Chidester, *The Long Sword Gloss of GNM Manuscript 3227a* (Somerville: HEMA Bookshelf, LLC, 2021), 33-34

9   Stephen Cheney, *Ringeck Danzig Lew (RDL) Longsword* (Coppell: Bucks Historical Fencing / Philadelphia European Martial Arts Collective, 2020), 8-9

10   Michael Chidester, *The Long Sword Gloss of GNM Manuscript 3227a* (Somerville: HEMA Bookshelf, LLC, 2021), 31-32

11   Michael Chidester, *The Long Sword Gloss of GNM Manuscript 3227a* (Somerville: HEMA Bookshelf, LLC, 2021), 53

12   Michael Chidester, *The Long Sword Gloss of GNM Manuscript 3227a* (Somerville: HEMA Bookshelf, LLC, 2021), 31-32

13   While its possible to perform the cut while passing forward, this can result in a bind where the fencers are too near one another to reliably perform the zornort. An alternative interpretation is to pass backward, away from the opponent's oberhau, while executing the zornhau. However, this may result in a wide measure which is not conducive to quickly shooting the point forward. The author's preferred footwork is a small lateral step with either the forward or rear foot.

14   There are a couple of common interpretations for this action. One method is winding the sword very high and very laterally. This possibly grants greater leverage, but it reduces reach and is quite slow if the fencer is making a lot of movement. The alternative is to make minimal upward motion with the hands while attempting to keep them forward. The latter interpretation is typically the most quick and efficient method.

15   Ringeck and Danzig do not describe any footwork for the schielhau when used to counter an oberhau. Lew however does describe jumping forward with the right foot.

16   Some modern practitioners attempt to perform the scheitelhau by bending their wrists while trying to cast the cut towards their target. Cutting in this manner is often performed while using the "push-pull" mechanics which this books discourages. The efficacy of this cut is debatable but performing the scheitelhau in this way is perhaps not "wrong" if the fencer can easily initiate the followup thrust from there.

17   Stephen Cheney, *Ringeck Danzig Lew (RDL) Longsword* (Coppell: Bucks Historical Fencing / Philadelphia European Martial Arts Collective, 2020), 180

18   This is somewhat speculative. The strike appears in Hans Talhoffer's 1467 treatise, though there isn't a gloss. Andre Lignitzer references the sturtzhau in his sword and buckler treatise and this is the primary inspiration for the interpretation presented in this book. In Lignitzer's fifth sword and buckler play, the fencer feints a high thrust which is then turned into a low thrust, delivered under the opponent's shield.

19   Descending false edge deflections like this show up throughout Meyer's writings, though he doesn't necessarily assign these actions a specific name. It could even be argued that Meyer would define this action as a krumphau, rather than a sturtzhau, as his primary prerequisite for the cut is the hands are crossed. Example of these false edge deflections can be found in Meyer's 1570 Einhorn plays. Jeffrey Forgeng, *The Art of Combat* (Yorkshire: Greenhill Books, 2006), 81-83

20   Even if the prellhau hits the target, it can very easily be redirected into another opening. So it can function like a feint, but it is arguable safer, as the attack is making contact with an opening. This is somewhat speculative, though the name "bounce strike" implies that part of its function is to spring away from contact.

21   Jeffrey Forgeng, *The Art of Combat* (Yorkshire: Greenhill Books, 2006), 92

CITATIONS

ART

Joachim Meyer. Illustration. iStock. 1570. https://www.istockphoto.com/vector/long-sword-fighting-by-joachim-meyer-published-in-1876-gm490375797-40176646

Talhoffer, Hans. Illustration. Wikimedia Commons. 1467. https://comons.wikimedia.org/wiki/File:De_Fechtbuch_Talhoffer_002.jpg

BOOKS

Jeffrey Forgeng, *The Art of Combat* (Yorkshire: Greenhill Books, 2006)

Michael Chidester, *The Long Sword Gloss of GNM Manuscript 3227a* (Somerville: HEMA Bookshelf, LLC, 2021)

Michael Edelson, *Cutting with the Medieval Sword: Theory and Application* (CreateSpace Independent Publishing Platform, 2017)

Richard Marsden, *Historical European Martial Arts in its Context* (Phoenix: Tyrant Industries, 2016)

Stephen Cheney, *Ringeck Danzig Lew (RDL) Longsword* (Coppell: Bucks Historical Fencing / Philadelphia European Martial Arts Collective, 2020)

WEBSITES

https://wiktenauer.com/

ACKNOWLEDGMENTS

Kyle Griswold began his HEMA career in 2011, under the tutelage of Richard Marsden. Benefiting from over a decade of law enforcement experience, Kyle has served in the capacity of a patrol deputy and SWAT operator. The valuable lessons learned through his past training and experiences has allowed Kyle to develop a practical approach to teaching HEMA.

Kyle is the co-founder and head instructor of Mordhau Historical Combat in Mesa, Arizona. When not teaching at his studio, Kyle is often running workshops and classes around the world. Kyle has extensive competitive experience and has medaled in a wide variety of weapons. Recently, Kyle was one of the fencers selected to represent the North American team at the 2019 Summer European Games.

Sean Franklin began his HEMA career in 2011, and hit the ground running. Due to his prior experience in the Canadian High-Performance Sport System he was able to apply his physical conditioning and disciplined training focus to develop as a martial artist at a rapid rate, being able to outfight many club head instructors after only a few years' experience. Sean has experience in many weapons and traditions, having competitive medals in most tournament weapon sets.

Sean's experience in sports coaching has allowed him to rapidly develop as a martial arts instructor, working to develop high level martial artists by instructing at schools and events around the world. Recently he served as a delegate and coach for the North American team at the 2019 European Summer Games (yes, you read 'North American team at the European Games' correctly.) A passionate advocate of test cutting, Sean finds the practice is useful for both preserving martial validity and as a tool to assist the development of quality body mechanics.

Brittany Reeves is an internationally recognized HEMA instructor, competitor, and coach. Taking her first HEMA class in 2011 in Vancouver, Canada, Brittany began a longterm hobby turned passion. She later co-founded Mordhau Historical Combat in 2018 where she is now the Head Instructor and coach.

In 2019, Brittany was selected to represent Canada and USA on Team North America for the European Games in Minsk, Belarus, and has medaled and instructed across North America and Europe. She holds medals in multiple categories including longsword, axe, wrestling, and cutting.

Brittany holds a Bachelor's Degree in Ancient and Medieval History, with a minor in Art History, from the University of Calgary in Alberta, Canada.

Thank you so much for your love and support, wife.

-Kyle

www.ingramcontent.com/pod-product-compliance
Lightning Source LLC
Chambersburg PA
CBHW042239140626
46547CB00036B/40